Table of Contents

1.BASICS

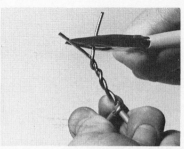

Most home electrical jobs require very little physical strength and no power tools or special skills—though a bit of practice may help in performing some tasks.

Most home electrical work is surprisingly easy to do; the system is logical; replacement parts are standardized so they will fit anywhere in the U.S.

Most home electrical work is safe to do if you understand and follow the few simple safety rules described in this book.

This book provides the step-by-step words and pictures—including safety precautions—for doing a wide variety of home electrical repairs and modifications.

A section starting on page 8 describes how electricity works and what electrical terms mean. If electrical work is new to you, read that section first.

From time to time you will notice hints from Practical Pete. Everybody makes mistakes sometimes—Practical Pete will help you avoid the more common ones and will give you tips on quicker or easier ways to get a job done.

The sizes, shapes, and characteristics of electrical materials are standardized, and the materials are relatively inexpensive. You can therefore buy what you need at modest cost and with complete confidence that—with the help of this book—you can do a safe, professional-type job that will increase the convenience, beauty, value, or safety of your home.

Most electrical work can be done by the homeowner without a permit and without conflict with local codes. However, requirements do vary. Contact your city or village building department for a copy of local regulations. Read the section in this book on the *National Electrical Code*.

Electrical work by the homeowner does not affect fire insurance coverage. However, if a fire loss claim results from wiring you installed, the company may classify you in a higher risk category.

Adequate precautions and instructions on checking your work are included in this book, beginning opposite.

The UL seal. Underwriters' Laboratories Inc. is a nonprofit organization that performs tests on electrical products. If an item meets minimum safety standards for avoidance of fire and shock, it is listed with the Underwriters' Laboratories. Listed items display the UL seal on the item itself or on its package. Note that the UL seal indicates minimum compliance with safety standards. Products having a wide range of prices may all have the UL seal. Higher-priced items will generally have additional features for convenience or longer life.

National Electrical Code Book. This book provides thorough, very detailed instructions for determining the adequacy of your wiring system and guides you in the safe repair and modification of your appliances, lighting, and wiring.

The chances are your house wiring was safe and adequate when your house was built. It is probable, too, that your house wiring was originally done in accordance with the National Electrical Code. This code was developed by the National Fire Protection Association and is periodically revised. The code itself has legal status only insofar as it has been adopted by cities, towns, counties, and other governmental units as part of local building codes and ordinances. The status of the code in your locality can be determined by contacting the building inspection department of your local government.

The National Electrical Code does not provide step-by-step instructions as this book does. Rather, the code specifies the preferred method of performing electrical wiring tasks and the proper materials to use in various applications. The instructions in this book conform to the recommendations of the National Electrical Code. Incidentally, the code does *not* limit or prohibit electrical work by a homeowner.

If you would like a copy of the code, write to: The National Fire Protection Association, 470 Atlantic Avenue, Boston, Mass. 02210. The current price is $5.50 postpaid. The code may also be available in your local library or from an electrical supplier.

ELECTRICAL FIXTURES, WIRING & APPLIANCES

ELECTRICAL FIXTURES, WIRING & APPLIANCES

Created and editorially produced for Petersen Publishing Company by Allen D. Bragdon Publishers, Inc., Home Guides Division

STAFF FOR THIS VOLUME:

Editorial Director	Allen D. Bragdon
Managing Editor	Michael Donner
Art Director	John B. Miller
Assistant Art Director	Lillian Nahmias
Text Editors	Allen D. Bragdon, Jayne Lathrop
Copy Editor	Jill Munves

Contributing Artists

Pat Lee, Clara Rosenbaum, Jerry Zimmerman

Contributing Photographers

Jack Abraham, Jayne Lathrop, Michael Mertz, John B. Miller

Cover design by	"For Art Sake" Inc.
Text originated by	Joseph H. Foley

Joseph H. Foley is Senior Publications Engineer for Sperry Rand Corporation. He has 30 years experience in writing instructions to make complex technical procedures clear and complete for people who know nothing about the subject. He does his own home rewiring successfully and, consequently, is much in demand by his neighbors.

ACKNOWLEDGEMENTS

The Editors wish to thank the following individuals and firms for their help in the preparation of this book: Anaheim Manufacturing, a Tappan division; Borger's Television; Circle F Industries; Fedders Corporation; General Electric Company; Gould, Inc., Electrical Products Group, Special Markets Department; Harvey Hubbell, Inc., Wiring Device Division; William C. Huber, Pass & Seymour, Inc.; Illuminating Engineering Society; Lightcraft, NuTone Division; Sperry Remington, Division of Sperry Rand Corporation; William M. Teed, Slater Electric Inc.; Westinghouse Electric Corporation, Lamp Division; The Wiremold Company; Whirlpool Corporation.

Petersen Publishing Company

R. E. Petersen/Chairman of the Board
F. R. Waingrow/President
Alan C. Hahn/Director, Market Development
James L. Krenek/Director, Purchasing
Louis Abbott/Production Manager
Erwin M. Rosen/Executive Editor,
Specialty Publications Division

Created by Allen D. Bragdon Publishers, Inc. Copyright © 1977 by Petersen Publishing Company, 8490 Sunset Blvd., Los Angeles, Calif. 90069. Phone (213) 657-5100. All rights reserved. No part of this book may be reproduced without written permission. Printed in U.S.A.

Library of Congress Catalog Card
No. LC 77-076140
Paperbound Edition:
0-8227-8002-X
Hardcover Edition:
0-8227-8014-3

The text and illustrations in this book have been carefully prepared and fully checked to assure safe and accurate procedures. However, neither the editors nor the publisher can warrant or guarantee that no inaccuracies have inadvertently been included. Similarly, no warranty or guarantee can be made against incorrect interpretation of the instructions by the user. The editors and publisher are not liable for damage or injury resulting from the use or misuse of information contained in this book.

Safety rules

1. Always turn power off before working on an electrical circuit. (See Service panel, page 16.)

Rule 1 applies to all jobs described in this book. There are no exceptions to this rule.

2. Always test before you touch. (The device you need to make the test is described on page 13. The step-by-step instructions tell you when to make the test.)

Rule 2 is necessary because it double-checks rule 1. Defects in house and appliance wiring can cause power to be present at times and places when it should not be.

Additional precautions

1. Always wear rubber-soled shoes.
2. When in your circuit breaker or fuse panel area, wear rubber gloves and avoid contact with damp floor by standing on a dry board or rubber mat.
3. Plan your job carefully in advance. Inspect the areas in your home in which you will be working. Note safety hazards and decide how to avoid them. Safety precautions for particular jobs are included in the detailed instructions in this book.
4. Plan to work as much as possible during daylight hours. Power will be off most of the time you are working. You may still need a flashlight for dark areas, but generally daylight will make your work easier and a great deal safer.
5. Don't hurry and don't work when you are tired. Haste and fatigue lead to carelessness. Carelessness leads to trouble.
6. Periodically test the circuit tester. Make sure the voltage tester (described under Basic tools) is working properly by checking it in a "live" receptacle.
7. Don't work alone. Have someone around for assistance if you need it.

Shock

What is electric shock? Electric shock occurs when a human body becomes a path along which electric energy can flow. When electric energy can flow along two or more paths, the bulk of the flow will occur along the path that offers the least resistance to the flow. The human body is generally a low-resistance path, and if it accidentally becomes part of an electrical circuit, it will experience the heaviest flow of electrical energy in that circuit.

How can it be avoided? The way to avoid shock, then, is to avoid becoming a path for the flow of electrical energy. The standard methods of avoiding shock are by insulation and grounding.

Insulation is the containment of electrical energy in wires, lamps, electrical outlets, appliances, etc., by enclosing the energy carrier in some material through which electrical energy cannot flow. The rubber or plastic covering on electric wire is insulation in this sense.

Grounding means simply providing a better (that is, lower resistance) path to ground for electric energy than the human body. (For a description of electrical grounding, refer to How electricity works, page 8.)

If shock occurs, the victim may not be able to release his grip on the "hot" lead. If main power can be turned off immediately, do it. If not, use a nonconductor such as dry wood or cloth to break the victim's grip on the hot lead. Don't touch him yourself.

When the victim is free of contact with power, call a physician or rescue squad. Keep the victim warm and give artificial respiration by any approved method until help arrives.

Who tests the tester?
A voltage tester is a handy little device to have around. It's just two wires with a little bulb between them that lights up when there is power in a circuit. Before I start working on an electrical fixture I poke the tips of the tester's wires into the slots of a wall plug, for example, or touch one end to the live wire on a switch and the other to the side of the metal box to make **sure** I flipped off the right switches and pulled the right fuses to turn off the power.

But what if the tester doesn't work? Have you checked your voltage tester in a **live** outlet recently?

Practical Pete

System grounding

Proper grounding of electrical circuits and appliances is the most important single safety feature in your home. If you are a bit rusty on what grounding means electrically, the section on How electricity works, beginning on page 8, will help refresh your memory in short order.

Codes and standard wiring practices require that all the metal boxes used for switches, outlets, and fixtures in your home must be grounded. A simple test to assure that this ground is not faulty is described on page 13.

The complete home electrical system or particular parts of it—especially outdoor wiring—can be protected by devices known as ground fault interrupters (GFIs). These devices turn off power to a circuit if current leakage develops in that circuit. Individual appliances can be grounded by connecting the outer metal enclosure of the appliance to any portion of the plumbing system—preferably the cold water line. Appliances that have three-pronged plugs are automatically grounded when plugged into a mating receptacle in a circuit with a properly wired grounding system. A three-pronged plug may be used with a grounding adapter that fits a two-slot outlet. This should be thought of as a temporary device until a grounding receptacle can be installed.

Ground fault interrupters

GFIs are supersensitive circuit breakers that monitor the current flowing in the black and white wires of a circuit. If no faults exist, the current in the two wires will be the same. If more current flows in one wire than the other, it means there is current leakage to ground. The GFI will sense this current difference and cut off power to the circuit within 1/50 of a second. This could save your life because a continuous flow through your body, even of one-third ampere for one-third of a heartbeat, can be lethal. The most recent electrical code requires GFIs in outdoor and bathroom circuits because current leakage can easily pass through a wet human body to the ground, and these are likely areas for such a mishap.

Three types are available: one for permanent installation, one that can be added to existing circuits by merely plugging it into an outlet, and a unit that combines the functions of circuit breaker and GFI in one package. The combined unit is designed for mounting in the circuit-breaker panel on the main service center.

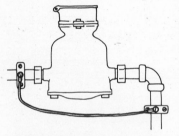

Make certain your water meter has a jumper wire. If any ground circuit is broken—even for a short time—the danger of shock is greatly increased. Many home electrical devices are grounded directly or indirectly to cold water pipes. Removal of the water meter may break this ground circuit. Check your meter to make certain a jumper is installed. If it is not, you can purchase the necessary clamps and cable in an electrical or plumbing-supply store. Install the jumper as shown. At the point where the cable touches the water pipe, scrape the pipe surfaces clean with a file or knife. Tighten clamps securely.

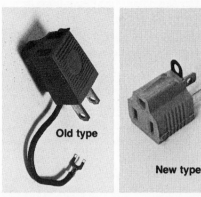

Old type

New type

A grounding adapter is for temporary use until the receptacle can be permanently wired into the grounding system.

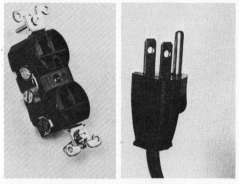

Permanently grounded receptacles have sockets that accept appliances wired with a three-pronged plug like this one.

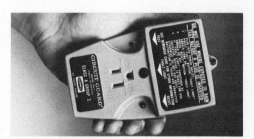

Portable GFI receptacle plugs into three-prong grounding outlets. First test the grounding in your wall outlet (see page 13).

Permanent GFI receptacle installed in place of a standard receptacle. It will work only if the separate grounding system in that circuit has been correctly wired.

Not all appliances should be grounded

In particular, this is true of appliances containing a heating element that can be touched by the user. This includes toasters, broilers, and electric heaters with exposed coils. The reason for not grounding these appliances is that the heating element is connected across the power lines when the unit is turned on. Grounding the outer shell or case of these appliances would provide a path for electric energy from the heating element to the outer shell. A user then touching the heating element and the outer shell—with a fork, for example—would be subject to severe shock.

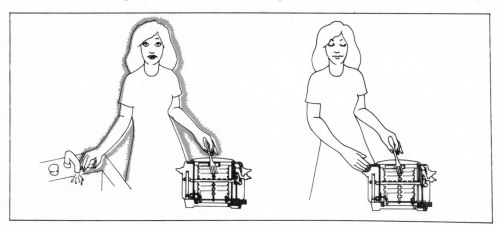

Why breakers trip or fuses blow when nothing's wrong

It may appear that circuit breakers go OFF or fuses blow in a haphazard manner. The reason for the overload is often a matter of timing. For the first few seconds after electric motors are started, they use four to six times as much power as they require when they are running. If a room air conditioner, large fan, or even a vacuum cleaner is on the same circuit with a TV set, electric heater, clothes dryer, etc., the temporary starting load of the motor in the air conditioner, fan, or cleaner added to the load already on the circuit from the other appliances will cause the breaker to go OFF or the fuse to blow. If you map your house circuits as described on page 19 of this book, you will be able to decide if you can avoid overloads by better use of your house circuits or if you need higher-capacity electric service installed.

Safeguards against fire

Electrical fires are caused by overloaded circuits or "short" circuits. The basic protection against overload and the danger of fire is provided by your house circuit breakers or fuse box. These devices are designed to turn off automatically part or all of the electricity if a serious overload and consequent overheating occurs anywhere in your home. Later on in this book you will learn how these devices work, what to do when they turn off your electric power, and how to deal with them safely.

To protect your house from electrical fire, the circuit breaker or fuse panel must be able to work as it was designed to work. Never restrict the action of circuit breakers, insert coins in fuse holders, or use oversized fuses. The safety provided by proper overload protection is more important than the temporary inconvenience of resetting circuit breakers or replacing fuses. This point cannot be overstressed.

How to toast a toaster
When I was a little kid, my mother nearly killed me when she caught me trying to fish stuck toast out of an electric toaster with a fork. I was lucky the toaster didn't do the job for her. Technically, if the heating elements are not grounded to the metal case, **and** you are not touching anything that is grounded (most things in a kitchen are), and you are lucky in a lot of other ways—you won't get electrocuted when you poke a metal object into a live toaster. But it's like trying to cross a busy highway with your eyes closed. Did you know, for example, that it is possible for old matted crumbs in a toaster to conduct electricity from the heating wires to the metal case?

Practical Pete

When should a licensed electrician do the job?

If a review of your home electrical system—as described in this book—shows the need for higher-capacity electric service, your utility company and a licensed electrician must do the work.

In some instances your local electrical code may require some types of work to be done only by a licensed electrician. In other cases the inspection requirements of your local code may make it desirable to have a licensed electrician do the job. Or, you may decide the particular job is too complicated or time-consuming for you to do. Even if you decide you would rather have a licensed electrician do the job, this book can be helpful. Read the sections of the book that cover the type of work you will have done. You will then be familiar enough with materials and work involved to discuss the job you want done and to judge the time and cost estimate. You may even get a lower price if you appear knowledgeable.

How electricity works

Working safely and efficiently with your home wiring and appliances is easier if you understand what electricity is and how it works. Basic electrical terms are briefly defined on the opposite page. If you want more than a definition, read the next few pages. They describe some of the how and why of electricity.

Electricity is energy

Electric utility generating stations convert either fossil fuel (coal, oil), hydroelectric energy (flowing water), or atomic energy (nuclear reactors) to electrical energy. The electric energy so generated is transported by wires to the factories, offices, schools, and homes that use it.

The basic building blocks of all matter are atoms. Electrons are one of the tiny particles that form atoms. Electrons are the stuff that electrical energy is made of. Some electrons can move from one atom to another. These are called "free" electrons. Metals such as copper, steel, and aluminum are called conductors because their atoms have many free electrons and so can conduct electricity efficiently. Wires made of copper, steel, or aluminum provide an ideal way to transport electrical energy with little loss of power.

The atoms that make up materials such as rubber, plastic, paper, and wood have almost no free electrons. These materials are called insulators because they cannot conduct electricity. A safe and efficient way to move electric energy, then, is to enclose a wire made of copper, steel, or aluminum in some insulating material and then use the wires to carry electricity from the generating plant to the final user.

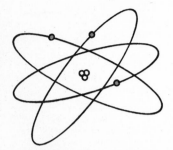

This diagrammatic model of an atom shows electrons (color) orbiting around the nucleus at the center. Electrical energy is made by making free electrons move in the same direction so they jump from one atom to the next.

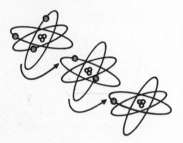

A free electron leaves the orbit of one atom, is temporarily "captured" into the orbit of the next, leaves that for a third, and so on. Metals with atoms having lots of free electrons, like copper, make good conductors.

Why are two wires needed to operate electrical devices?

Electricity is generated by causing all the free electrons in a conductor to move in the same direction. This creates a surplus of electrons in the atoms of one wire at the output of a generator and a shortage of electrons in the atoms of a second output wire from the generator. When an electrical device is connected to these two wires, electrons will move along an electrical path through the device in order to restore the natural balance. As long as the generating station at the source continues to operate, the shortage and surplus in the two wires will be maintained and electron movement will continue. The phrase "current flow" is used to describe this electron movement. The rate of the current flow (that is, the number of electrons that pass a point in one second) is measured in units called amperes, or more commonly, amps. The device that measures this current flow is called an ammeter.

The pressure that exists to restore the electron balance depends upon how large the difference is between the surplus and the shortage. The greater the difference, the higher the pressure. This pressure is called voltage and the units in which it is measured are called volts. The device for measuring voltage is called a voltmeter.

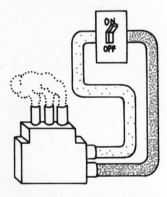

A generator creates a surplus of free electrons in the atoms of one wire and a shortage in another. When the two wires are connected (when you flip a switch to ON, for example) the excess electrons rush through toward the atoms where there is a shortage, creating electrical current.

Why are some wires called *grounding, neutral* or *hot*?

Ground means simply Mother Earth or something connected to earth, such as a cold water pipe in your home or a copper rod driven onto the ground outside your house near where electric power enters.

The earth is such a huge volume of matter that a measurable surplus or shortage of electrons never exists. Earth or ground, therefore, is always electrically neutral. Ground, and wires connected to ground, can accept electrons or give them up as necessary to cause current to flow between ground and a point at which a shortage or surplus exists.

While both grounding wires and neutral wires are connected to ground, there is a difference in the job each performs in electric wiring. The job of the ground wire is to provide a path to ground for electric energy when faults occur in the primary power wiring or in electrical devices. Throughout this book the term *grounding wire* refers to these safety wires. Grounding wires may have green insulation or be bare (no insulation). Throughout this book the term *neutral wire* always means the primary power wire with white insulation. The job of the white wire is to provide the normal path for return current flow to the source when no wiring faults exist. Throughout this book the term *hot wire* refers to the wire with black or red insulation. This is the wire that causes current to flow between it and the neutral wire (or grounding wire if a fault occurs).

Normally the hot wire is more dangerous because it can cause current to flow between it and *any* path to ground. In some cases in home wiring, wires with black, red, or white insulation are used for some other-than-normal function all or part of the time. When this is done, the wire should be marked at each end and at any terminal point with the color of its "live" function; that is, either black or red. As this marking may be forgotten or may wear off or flake off with time, NEVER rely on color of insulation alone to tell you a wire's function. *Always test before you touch.*

Definition of electrical terms

Alternating current (AC). The form of electric power used in almost all home utility systems. In this form of electric energy the voltage and current are constantly changing in amount and periodically changing in direction of flow.

Amps (amperage). The amount of current that flows in a circuit as a result of the voltage applied and the resistance of the load (appliance, light bulb, etc.).

Ballast. Part of a fluorescent fixture that controls voltage.

Branch circuit. Each circuit wired from the service panel to various parts of your house. Each branch circuit has its own circuit breaker or fuse for overload protection.

Circuit. A combination of source, conductor, load, and switching that enables electric power to do some useful work.

Circuit breaker. An automatic switch that cuts off power to a circuit when the current flow exceeds the rating of the circuit breaker. Circuit breakers can also be operated by hand to turn off power when desired (such as while making a repair).

Conductor. Any wire, bar, or strip of metal that offers little resistance to electric current and is therefore used for carrying it.

Direct current (DC). This is the form of electric energy available from batteries. If the battery source remains charged, the voltage in a DC circuit remains constant.

Electron. One of the tiny particles that make up atoms. Some electrons are called "free" because they can move from atom to atom and produce an electrical current.

Fuse. A safety device that breaks the flow of current when the amperage in the circuit exceeds the rating of the fuse.

Grounding wire. This is a safety wire. It is usually a bare wire but may have green or green and yellow insulation. This wire provides a safe path to ground for hot wire current, in the event of a circuit fault.

Hertz (Hz). A term that is replacing the older cycles-per-second. Hertz applies to AC power only. The number of Hertz of an AC source is the number of times per second the electric power goes through a complete change in amount in each direction. Most home power systems are 60 cycles-per-second or 60 Hertz.

Hot wire. The electric wire in your home usually having black or red insulation. This wire can cause current to flow through any path to ground. Though any wire can give a shock, a hot wire is most dangerous.

Kilowatt hours (KWH). The unit to measure power consumed in your home. One KWH equals 1,000 watts used for one hour. Your electric bill is calculated in kilowatt hours.

Load. The operating device in an electric circuit that converts electric energy to some other form; light in a bulb, mechanical movement in a fan, heat in a toaster, etc.

Neutral wire. The electric wire in your home having white insulation. This wire provides the normal return path for current flow to the power source.

Ohm. Unit of measurement for resistance to current flow.

Overload. A condition that results from too many devices on one circuit or from a circuit fault. Overloads can cause overheating in wires and tripped circuit breakers or blown fuses.

Power, electrical. The product of voltage and current that provides energy to do work. The unit of measurement of power is the watt.

Resistance. The characteristic of some materials and devices that restricts the flow of current. Unit of measurement is the ohm.

Service panel. The unit that distributes power to the various circuits in your home. Either circuit breakers or fuses are located on this panel.

Source. The point at which power originates. In your home, the service panel may be considered the source.

Starter. An automatic switch that controls current flow in a fluorescent fixture.

Switch. A device that can interrupt and restore current flow in a circuit.

Transformer. In AC power systems, transformers are used to raise or lower voltage.

Volts (voltage). The pressure that causes current to flow between a hot wire and a neutral or ground wire. In home electrical systems the voltage is either 120 volts or 240 volts, depending on type of wiring.

Watts (wattage). A unit of measure of the power consumed by a load. The wattage is calculated by multiplying the voltage by the amperage. For example, if a current of two amperes flows in a 120-volt circuit, the load is consuming 240 watts.

Current is measured in amperes. An amp is a measure of the number of free electrons moving past a point. (If dust particles being sucked through a hose into a vacuum cleaner were electrons, you would have to count 6.28 billion billion per second to equal one amp.)

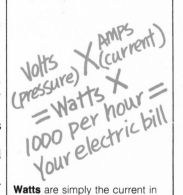

Voltage is the amount of pressure pulling the electrons through the conductor (like the suction power in the hose). The resistance to current flow caused by the load uses up the pressure (volts) but doesn't change the number of electrons (amps) flowing through the load and back into the circuit. Most houses are wired to supply two voltages—about 120 volts for most outlets, and about 240 for loads that require a lot of power.

Volts (pressure) X AMPS (current) = Watts ÷ 1000 per hour = Your electric bill

Watts are simply the current in amps multiplied by the pressure in volts. Watts indicate the rate at which the load is consuming the power you pay for on your electric bill. A radio uses about 10 watts, an electric frying pan about 1,000, a clothes dryer about 5,500. (This is one reason why modern appliances make us use so much more power per capita than we used to—and why we face the shortages in energy needed to generate it.)

What is an electric circuit?

A closed path of electron movement is required to put electric energy to use. This closed path is called a circuit. Every circuit has four parts:

1. Source

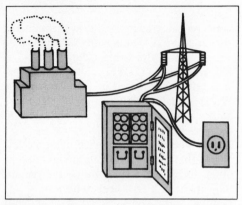

The true source is the utility generator. However, for practical purposes, the source may be thought of as the circuit breaker panel or fuse box from which power is distributed throughout your home. Junction boxes and wall outlets may be thought of as secondary sources.

2. Conductors

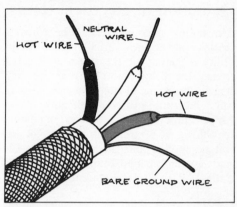

Conductors are the wires that carry the electric energy to the point at which it will be used. Conductors offer little (but not zero) resistance to current flow because the atoms in the metals they are made of have lots of free electrons and therefore can transport it efficiently.

3. Load

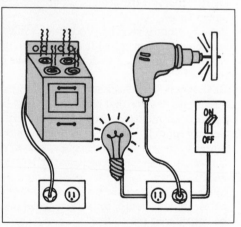

The load is any device (a light bulb, your toaster or your washing machine, for example) that uses electric energy to perform some work. The load (unlike the conductors) offers resistance to current flow. This resistance makes it possible for the device to convert electric energy to another form (heat, light, mechanical movement, etc.). The load resistance determines how much current will flow in a circuit. Increase the resistance in the load and you automatically decrease the current flowing in the circuit. If the voltage applied to a circuit is multiplied by the current in amperes, the result is in units called watts. The watt is a unit of power and indicates the rate at which the load is consuming power.

4. Switches

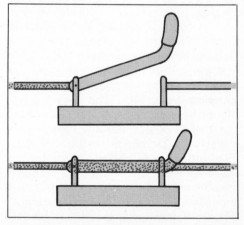

The fourth element in a circuit is a means of controlling the energy. This is a device that can interrupt and restore current flow as desired by the user. Switches used in home electric systems control current flow by inserting a very high resistance (air gap) in the circuit to stop the current flow and by removing the resistance (closing the air gap) to start current flowing again. A basic principle of electrical wiring is that switches should always be wired into the hot (black or red) line leading directly to a device or outlet. When switches in the hot line are turned off, no hot line power is present in the load device. This greatly reduces the possibility of injury or damage resulting from accidental grounding.

Important: Turn off power at main panel before working on switches or outlets. Even when a hot line switch is off, one terminal on the switch is still connected to the power source. Before doing any work on the switch, the power source must be turned off by setting a circuit breaker to OFF or removing a fuse.

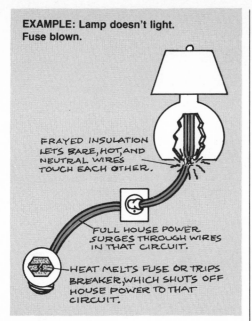

EXAMPLE: Lamp doesn't light.
Fuse blown.

FRAYED INSULATION LETS BARE, HOT, AND NEUTRAL WIRES TOUCH EACH OTHER.

FULL HOUSE POWER SURGES THROUGH WIRES IN THAT CIRCUIT.

HEAT MELTS FUSE OR TRIPS BREAKER, WHICH SHUTS OFF HOUSE POWER TO THAT CIRCUIT.

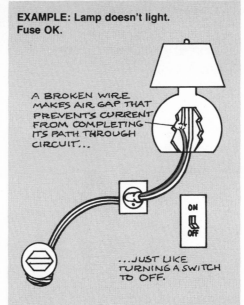

EXAMPLE: Lamp doesn't light.
Fuse OK.

A BROKEN WIRE MAKES AIR GAP THAT PREVENTS CURRENT FROM COMPLETING ITS PATH THROUGH CIRCUIT...

...JUST LIKE TURNING A SWITCH TO OFF.

Short circuit

A short circuit is a fault that occurs when a low resistance path exists between the hot lead and some grounded points. When this happens heavy current will flow in the circuit for as long as it takes for the circuit breaker to trip to OFF or for the fuse to blow. Because the circuit breaker or fuse cuts off the current flow quickly, the circuit wiring will not have time to overheat to the point where fire may occur.

Open Circuit

An open circuit is a fault that cuts off current flow in a circuit. The break that causes the open circuit may be in the hot line, the neutral line, or the load. An open circuit appears to be off. The break that causes an open circuit may be quite small. Vibration or even changes in temperature may suddenly turn the circuit on again. When looking for the cause of an open circuit, be sure power is off at the source.

What do AC and DC mean?

The electrical energy that flows in circuits and operates loads can be generated in either of two quite different forms. One form is called alternating current (AC); the other form is called direct current (DC).

Alternating current is almost universally used for home electric power and is, therefore, the kind this book is primarily concerned with. In an AC circuit, the amount of voltage applied to the circuit is constantly changing from zero to a maximum and back to zero in one direction and then from zero to maximum and back to zero in the other direction. The maximum voltage is set by the generating plant. Because voltage is the pressure that causes current to flow, the current will also change from zero to maximum to zero and will reverse direction and repeat. The maximum amount of current, however, is determined by the load resistance and can vary as the load resistance varies. Each complete change from zero to maximum to zero in one direction and then zero to maximum to zero in the opposite direction is called one hertz (formerly cycle). The term hertz implies

"per second." So, 60 hertz means the same as 60 cycles per second. Hertz is abbreviated Hz. Cycles-per-second, which you will still see marked in some electrical devices, is abbreviated cps.

Direct current is most commonly found in homes in the form of electrical energy stored in batteries. In a DC circuit, the amount of voltage and the direction of application are constant. The amount of voltage is determined by the type and size of battery. The direction of current flow is also constant and, as in AC circuits, the amount of current flow is determined by the load resistance. Batteries convert chemical energy to electrical energy. The chemical energy can be in wet form, as in your car battery, or in dry form as in flashlight and transistor-radio batteries. Some batteries are designed to be recharged from an AC source. The voltage from all batteries, unless recharged, will gradually decrease. AC power can be converted to DC power for some uses in the home. The conversion is performed by a device called a rectifier or current converter.

To review the main points:

- Electricity is energy.
- Electricity consists of voltage (pressure) and amperage (electron flow).
- Electricity can be generated at a distant point and carried by wires to where it will be used.
- Electricity is used by applying it to a circuit. The circuit consists of wires connected to a source, a load connected between the wires, and one or more switches to turn the circuit on and off.
- Resistance determines the amount of current flow in a circuit. Conductors offer little resistance to current flow, loads offer moderate resistance, open switches are an extremely high resistance, and closed switches almost zero resistance to the flow.

Tools and supplies

With a few exceptions, you probably have the tools you need for most electrical repairs. All you need are common hand tools, some power tools, electrical supplies, and the special items described on the opposite page.

Basic tools

Long-nose pliers
Linemen's pliers
Screwdriver set (small and medium sizes will do the job)
Phillips-head screwdriver set (needed for appliance work)
Folding rule or steel tape
Utility knife or pocketknife
Hammer
Keyhole saw
Hacksaw
Flashlight

Supplies

Solderless connectors (get a variety of sizes and types)
Plastic electrical tape
Cable staples
Rosin-core solder (marked as suitable for electrical work)
Noncorrosive flux

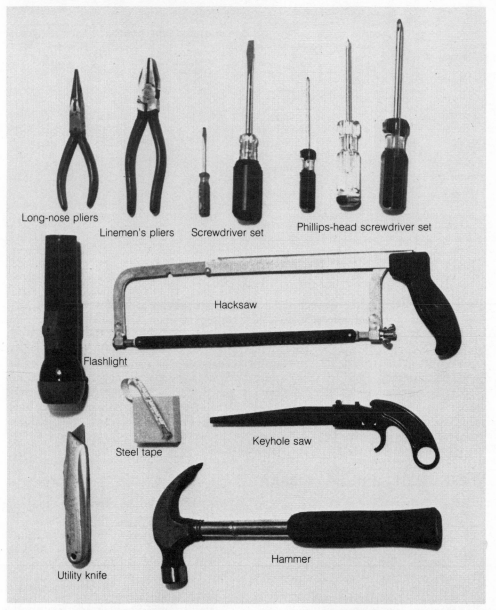

Long-nose pliers Linemen's pliers Screwdriver set Phillips-head screwdriver set

Hacksaw

Flashlight

Steel tape Keyhole saw

Utility knife Hammer

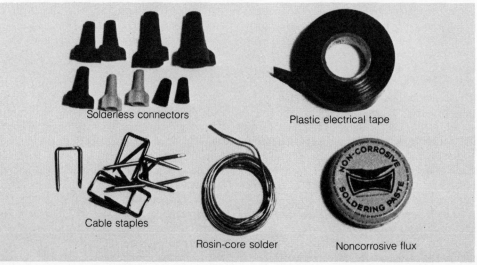

Solderless connectors Plastic electrical tape

Cable staples Rosin-core solder Noncorrosive flux

Specialized tools

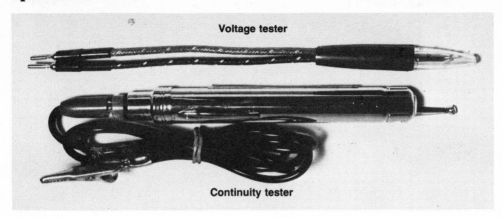

Voltage tester

Continuity tester

Voltage tester. This tester consists of a holder containing a neon bulb. Two probes are attached to the holder. The neon bulb will light when the probes touch the hot and neutral power lines or anything connected to those lines when power is present on the wires. For the procedures in this book, the tester is primarily used to make certain NO voltage is present before you touch any wiring or device. When no voltage is present the bulb does not light. Since the bulb will also not light if the tester is defective in any way, it is especially important that you test the tester from time-to-time. Simply place the test probes in a live receptacle. If the bulb lights the tester is OK. Make sure the tester you buy can be used on both 120 and 240 volt lines.

Continuity tester. This is a pen-like probe with an alligator clip lead attached. The probe contains a battery and bulb. When current flows from the alligator clip to the tip of the probe, the bulb lights. The continuity tester is always used with power off. The low battery voltage can be used to check switches, lamps, fuses and wiring. This is an excellent device to use to check your work before applying power for the first time. The continuity tester is easily checked before use by touching the alligator clip to the tip of the the probe. If the light goes on, the battery is OK and the tester is ready for use.

Testers
Two simple and inexpensive testers are absolutely essential to electrical work. If you own a handitester or similar volt-ohm meter, that will do the same job.

Testing correct grounding with a voltage tester

3-slot receptacle. Touch one probe inside the ground slot and the other probe to each of the prong slots in turn. The bulb should light in one of them.

2-slot receptacle. Touch one probe to the screw on the outside of the cover plate and the other probe in each slot in turn. The bulb should light in one of them.

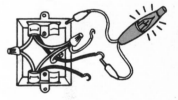

Switch. Unscrew the cover plate and remove the switch without touching any wires. Touch one probe to the metal box (or white ground wires on a plastic box). Touch the other probe to the bare end of the "hot" black wires in turn. The bulb should light when the probe is touched to the one which leads back to the main service panel, not to the other one.

Testing for dead circuit

Receptacles and switches. To be sure power is off, insert the probes into the two slots of an outlet. For switches, follow the procedure above for testing correct ground. If power is off, the bulb will *not* light. The procedure for testing wall or ceiling fixtures and outlets is described from page 48 on.

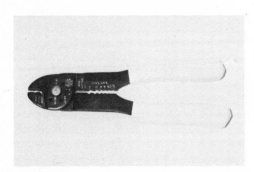

Wire stripper. Many types of tools are available for stripping insulation from wire. All consist of a plierslike tool with cutouts on the jaws corresponding to various wire sizes. The cutouts allow the stripper to cut through the insulation without cutting or nicking the conductor. Some strippers also have provision for cutting wires and small bolts.

Fuse puller. If there are cartridge fuses on your service panel, you will need a fuse puller for safe removal of fuses. Check the sizes of the fuses you will have to remove before you purchase a fuse puller. Make certain the puller you buy is the right size for your use.

Fish tape. This is a flexible tape available in various lengths. It is used to "fish" wire through walls and floors when installing new wiring. The tape has a hook at the end to which the wire can be attached after the tape is worked through the opening. The tape is then withdrawn pulling the wire through.

Soldering gun. If your work involves wire splicing, it is desirable to solder the splice to assure good electrical contact. A high heat electric iron, or gun, or a pencil flame propane torch, will heat the joint faster and assure a good flow of solder.

Power to your house

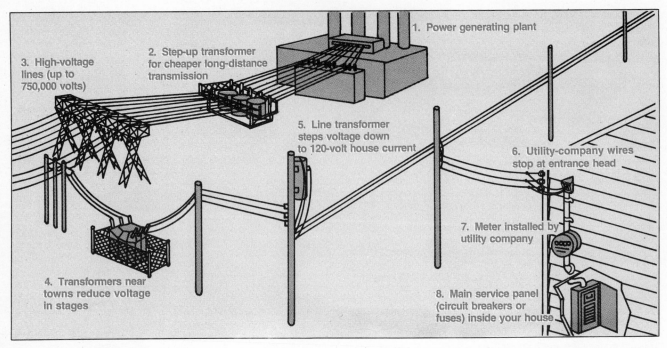

1. Power generating plant
2. Step-up transformer for cheaper long-distance transmission
3. High-voltage lines (up to 750,000 volts)
4. Transformers near towns reduce voltage in stages
5. Line transformer steps voltage down to 120-volt house current
6. Utility-company wires stop at entrance head
7. Meter installed by utility company
8. Main service panel (circuit breakers or fuses) inside your house

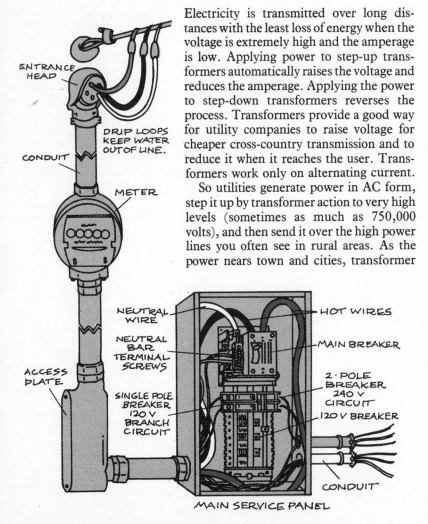

ENTRANCE HEAD

DRIP LOOPS KEEP WATER OUT OF LINE.

CONDUIT

METER

ACCESS PLATE

NEUTRAL WIRE

NEUTRAL BAR TERMINAL SCREWS

SINGLE POLE BREAKER 120 V BRANCH CIRCUIT

HOT WIRES

MAIN BREAKER

2-POLE BREAKER 240 V CIRCUIT

120 V BREAKER

CONDUIT

MAIN SERVICE PANEL

Electricity is transmitted over long distances with the least loss of energy when the voltage is extremely high and the amperage is low. Applying power to step-up transformers automatically raises the voltage and reduces the amperage. Applying the power to step-down transformers reverses the process. Transformers provide a good way for utility companies to raise voltage for cheaper cross-country transmission and to reduce it when it reaches the user. Transformers work only on alternating current.

So utilities generate power in AC form, step it up by transformer action to very high levels (sometimes as much as 750,000 volts), and then send it over the high power lines you often see in rural areas. As the power nears town and cities, transformer action is again used to reduce the voltage. The reduction is done in steps rather than all at once, so that the highest practical voltage level can be maintained for as long as possible. The final voltage reduction takes place on the line transformers you see mounted on utility poles in your neighborhood. In many communities power lines are underground. Of course, neither the lines nor the transformers are visible in these areas.

Unless your house was built before World War II, the chances are that power enters your house on three lines. Two of these are hot lines and the third is a neutral line. This combination provides two voltage levels for use in your home: 120 volts and 240 volts.

Various devices are used by utility companies to secure the power lines to the house. Power lines must be able to withstand heavy winds, ice loads, etc. In many cases the neutral wire is made to do double duty for this purpose. The wire is made of braided steel strands for strength and is thicker and stronger than the hot wires. The three wires are then held together by twisting or enclosing them in an outer sheath so that the anchoring of the neutral wire to the house provides support for the other two lines as well. The incoming power lines—heavily insulated or enclosed in a metal conduit—are then routed to your meter. Meters are almost always mounted on the outside of the house.

How to read your meter

If you know how to read your meter, you can check the reading against the amount on your bill from time to time and make certain the meter was read or estimated accurately.

Meters record the power used in your home in units called kilowatt hours. Watts, you will recall, are units of power equal to the voltage times the current. The prefix kilo means 1000, and a kilowatt hour therefore represents your use of one thousand watts for one hour. For example, a washing machine and a refrigerator running for one hour would consume about one kilowatt of electricity together.

Your meter reading is determined by the positions of pointers on five dials. Each dial has ten markings, from zero to nine. But the markings on different dials represent different amounts: units, tens, hundreds, and so on. The pointer always moves from 0 to 1 to 2 to 3, etc., and back to zero. However, the dials are marked in different directions. That is, the numbering of the 10,000 unit dial increases in a clockwise direction, the 1000 unit dial counterclockwise, etc.

Determine the direction of a dial rotation by the way the dials are numbered. All meters are not the same. As long as any electricity is being used in your home all the dials will be moving. The amount of movement will be ten times slower on each dial from right to left. Unless power consumption is unusually heavy, movement will be noticeable only on the right-hand dial. To read the meter, note the number the pointer has just passed—keeping in mind the direction of rotation—on each dial, starting from the left.

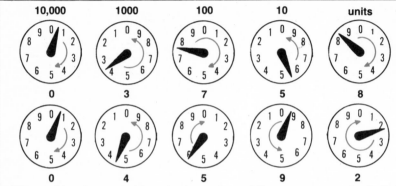

10,000	1000	100	10	units
0	3	7	5	8
0	4	5	9	2

If the readings above were taken about a month apart, the power consumption for that month would be the difference between the two readings—or 834 kilowatts.

For example: At left is a reading of 03758 or 3758 KWH. Meter dials are not precisely marked. A pointer on any dial may appear to be exactly on a number, but actually be above or below it. To decide which it is, note the pointer on the dial to the right of it. If the pointer to the right has not reached zero, the preceding dial has not reached the nearest number. If the pointer to the right has passed zero, use the next higher number for the preceding dial.

In the example at left, the middle dial appears to be on 6, but because the next dial to the right is below zero, 5 is the correct reading for the middle dial. Full reading 04592, or 4592 KWH.

Consuming watts

This list of appliance wattage ratings can be used to estimate the total power load in your home. You can also use it along with your circuit map to figure out how to distribute your appliance load more evenly over available circuits. The wattages given here are typical and should be satisfactory for estimating purposes. You can determine the exact wattage of your appliances from information on the manufacturer's nameplate. If the nameplate lists amperage only, multiply by the listed operating voltage (either 120 or 240 volts) to get the approximate wattage. If motor horsepower (HP) is listed, figure about 1,000 watts for one horsepower and gauge other sizes proportionately. For example, ½ or 0.5 HP = 500 watts. But for motors larger than 1½ HP, power consumption should be obtained from the manufacturer's published data. In addition to the wattages listed here, be sure to include all your lighting fixtures. Wattages are marked on bulbs.

Power ratings for typical home appliances

Appliance	Wattage	Appliance	Wattage	Appliance	Wattage
Air conditioner (room size)	800-1500	Fan, attic	400	Radio (solid state, plug-in)	10
Air conditioner (central, 240 v.)	5000	Fan, kitchen	70	Range (240 v., one unit)	8000-16,000
Blanket (single)	150	Fan, large pedestal type	500	Range (surface burners)	5000
Blanket (dual)	450	Fan, small table type	80	Range (oven)	4500
Blender	250-450	Floor polisher	300	Refrigerator	300
Broiler	1500	Freezer	600	Rotisserie	1400
Can opener	100	Frying pan	1000	Saw, radial	1500
Clothes iron (large unit)	1800-3000	Furnace (gas)	800	Saw, table	600
Clothes dryer (120 v.)	1400	Furnace (oil)	600-1200	Soldering iron	150
Clothes dryer (240 v.)	5500	Garbage disposal	900	Stereo hi-fi	300-500
Coffeemaker/percolator	600-750	Hair dryer	400	Sun lamp	250-400
Crock pot (2-quart)	75-150	Heater, portable	1300	TV, color (25″, solid state)	250
Deep fryer	1350	Heater, built-in	2000	TV, b & w (12″, solid state)	50
Dehumidifier	400-600	Hot water heater (240 v.)	2500	Toaster	1200
Dishwasher (with water heater)	1800	Iron	1000	Vacuum cleaner	600
Drill, portable	200-400	Microwave oven	650	Waffle iron	1100
Drill press	500	Mixer	150	Washing machine	900

Service panel

Fuse panel

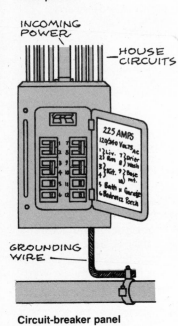

Circuit-breaker panel

From the meter, the three power lines enter your home and are connected to a service panel that divides the incoming power into branches or circuits. The service panel also provides overload protection by means of circuit breakers or fuses, which shut off the power when an overload is imminent.

Both types of service panel contain a means of shutting off all incoming power. On circuit-breaker panels, two main circuit breakers are provided. Fuse panels may either contain a main fuse block that can be removed or will have a separate box with a main power switch that must be turned off before the box can be opened. The box contains the cartridge fuses that protect the main line. In addition to providing a shut-off point, the main fuse or circuit breaker sets the maximum power available to your home. You should become familiar with the main shut-off on your service panel. In an emergency it is the best means of making certain that all power is off.

Both circuit breakers and fuses are rated for various loads, such as 15, 20, 30 amperes and higher. The rating of each circuit breaker or fuse is marked on your service panel, or on the breakers and fuses, or both. You can read the rating without touching the panel or turning off power.

If your home has three-wire service, you can easily determine the total power available. Note the rating of the circuit breaker or fuse on each side of the main input line. Add these two numbers. Multiply by 240. The result will be the total watts available in your home at any one time. For example, two 50-ampere breakers or fuses will allow 24,000 watts to be used.

By referring to the list of typical wattages for home appliances (page 15), you can determine the total wattage used in your home. If the total wattage used is 80 percent or more of the total available, a bit of additional calculation should be made. Keep in mind that the total wattage on your list would occur only if all appliances were on at the same time. Make a separate list of those appliances that are used only during one season of the year. (Air conditioners will not be used with the furnace or heaters, for example.) Include in your grand total only the appliance with the higher wattage rating. If this still does not give you a margin of safety greater than 20 percent, you should consider increasing your basic service. Changes in your basic power service must be done by your utility company. The 20 percent safety margin is a rough guide rather than an exact number. It is based on three considerations:

1. Electrical systems and devices age the same as other parts of the house. It's unwise to push them to the limit.

2. Even after the seasonal adjustment is made, the appliances in use at any one time are not uniform in their power needs. Appliances containing motors consume from four to six times as much current during the first few seconds of operation as they do when running at normal speed. The heavy starting current is caused by the need to overcome the inertia of the load the motor is driving. A reasonable margin of safety will prevent the starting current surge from tripping circuit breakers or blowing fuses.

3. Even with careful use and conservation of power, family electric needs tend to increase over the years. Be sure to allow for this increase when deciding on the service you require. Incidentally, the cost of increasing your electric service can prove to be a good investment even if you intend to sell your house. Appraisers give considerable weight to adequate electric service.

Household circuits

The service panel—whether circuit breakers or fuses are used—is the point at which the main power into your home is divided into individual circuits. Each circuit is separately protected by its own circuit breaker or fuse. Modern service installations have anywhere from 12 to 32 individual circuits. Four types of individual circuits are used in home power systems:

1. General purpose circuits. These circuits provide basic lighting and wall-outlet power. They are usually protected by a 15-ampere circuit breaker or fuse.

2. Appliance circuits. Kitchen and laundry areas generally have greater power needs, so these circuits are protected by 20-ampere circuit breakers or fuses.

3. Special-purpose circuits. These circuits serve a single large appliance—such as a furnace or washing machine—through a single, three-wire wall outlet. They are protected for a 20- or 25-ampere load.

4. 240-volt circuits. Heavy-load appliances—such as central air conditioners, ranges, and clothes dryers—are more efficiently operated at 240 volts than at 120. These are three-wire circuits on paired circuit breakers or fuses.

Circuit breakers

Circuit breakers are special-purpose toggle or push-type switches. These switches are triggered automatically but can also be operated manually. Circuit breakers are available in ratings of 15 to 150 amperes. The load rating is determined by the internal switch mechanism and is marked on the outside of the circuit breaker. Circuit breakers will remain in the ON position indefinitely if the load remains at or below the rating. If the load exceeds the rated amperage, the circuit breaker will automatically switch to the OFF or TRIPPED position before any damage can occur.

Push-button circuit breakers have an indicator to show whether the circuit is ON or OFF. To reset a push-button type, simply depress and release it. The indicator will change from OFF to ON. Toggle-type circuit breakers have either two or three positions. The condition of the circuit breaker is shown by the position of the toggle: up for ON and down for OFF. Three-position toggles have a center position marked TRIP. Toggle-type circuit breakers are reset by moving the toggle back to the ON position. Some three-position types must be moved from the center TRIP position to the down OFF position and then back to ON.

Circuit breakers normally are reliable devices and have a long life span. However, they can, and sometimes do, become defective. Replacing circuit breakers can be hazardous. It is recommended that you have a licensed electrician do the job.

If you want to replace your screw-in fuses with circuit breakers, you can buy individual screw-in circuit breakers and substitute them for the fuses yourself. Replacing an entire fuse panel with a breaker panel should be done by a licensed electrician. Be sure you match the voltage and amperage ratings of the new breakers to the existing fuses.

Fuses

All fuses contain a special piece of metal designed to melt when more than a specified amount of current flows through it. The type of metal chosen and the thickness of it determine how much current it can carry. The metal is enclosed in insulating material. All except cartridge types can be removed by hand.

Plug-type fuses have screw-in bases like light bulbs and can be screwed into the fuse panel in a similar way. The metal strip that protects the circuit is visible through a plastic window. When a short circuit blows a fuse, the plastic window is clouded by the flash. A blown fuse with a clear window usually means the circuit is overloaded.

Time-delay fuses are similar in appearance to the ordinary plug type, but these fuses are designed to carry more than the rated current for a short period of time. These fuses are particularly useful in circuits in which the load consists of a motor-driven appliance. The normal high starting current of the motor will not blow these fuses.

Type-S fuses (also called nontamperable fuses) are time-delay fuses with a special, separate base that is inserted into the fuse panel socket. The special base allows only one size fuse to be screwed into a particular socket. A 20-ampere fuse will not fit a 15-ampere socket. This prevents using a higher-rated fuse than the circuit allows.

When darkness falls
Suddenly everything goes black. Did the electric company go out of business? Did a tree fall on a power line? Did a fuse blow? I need a flashlight to see the fuse panel in the dark, but the kids lost mine on the last Boy Scout hike. Tomorrow morning I'll buy an inexpensive pencil flashlight and clip it to the handle of the fuse box.

Practical Pete

Cartridge fuses are used in heavy-load circuits or in main circuits that carry heavy loads. Both types make electrical connection by snapping into clips.

Ferrule-type cartridges (from 10 to 60 amperes) have round ends that make electrical contact.

Knife-blade-type cartridges (above 60 amperes) have flat ends for electrical contact.

Safety rules for fuse-panel work

All the general safety rules mentioned at the beginning of this chapter should be observed when changing fuses. In addition, special care must be taken because hot terminals are exposed on fuse panels.

1. Wear rubber-soled shoes. Stand on a dry board or rubber mat. Never stand on a bare, damp floor.
2. Use only one hand to insert or remove fuses. Keep the other hand in your pocket, or make certain you are not touching any surface that may be grounded.
3. Never place your finger or a hand tool in a fuse socket.
4. Never use a hand tool to bend or straighten a cartridge fuse clip.
5. Always use a fuse puller to remove cartridge fuses.

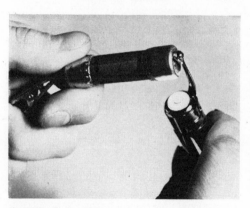

If an appliance fails on a circuit protected by a cartridge-type fuse, there is no way to tell whether the fuse is blown or is good just by looking at it. If you think it may have blown, throw the main cutoff lever to OFF and remove the fuse with a fuse puller. If the bulb in the continuity tester lights up when you simultaneously touch the caps at each end of the fuse, it is still good; the appliance may be faulty.

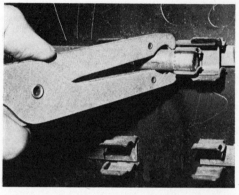

You must use a fuse puller to remove cartridge fuses because the two exposed clips that hold them tightly at either end carry a live voltage charge. In addition to the danger of severe shock if you touch them without having turned off the safety switch (the door of most cartridge-type fuse boxes won't open unless you first throw the switch), the cartridge is likely to be hot to the touch.

Why did the fuse blow?

Blown fuses

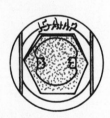

Short circuit usually causes discoloration of the plastic window.

Circuit overload melts strip slowly, so the plastic window usually stays clear.

Before resetting a circuit breaker or replacing a fuse, find out what caused the breaker to trip or the fuse to blow. In many cases you will know the cause of the overload because the circuit went off just as some appliance was plugged in or turned on. The section on mapping your house circuits, opposite, gives you some hints on how to make best use of the power you have available so that this can be prevented.

You can uncover other clues to the cause of an overload by touch and smell. If a defective lamp or appliance caused the overload, the cord and plug will be hot to the touch, and frequently a somewhat acrid odor can be detected from hot or burned insulation. But if there is no apparent sign of what caused the breaker to trip or the fuse to blow, the following procedure should solve the problem.

1. Turn off all the lights and disconnect all plug-in devices on the circuit. If you have a circuit map, this will be easy. If you do not have a circuit map and are not sure whether a device is on the overloaded circuit, assume it is on the bad circuit and turn it off or unplug it.
2. Remember: even after you have turned off and disconnected all the items on the circuit, the short that caused the overload may still exist in the wiring.
3. With a flashlight and a replacement fuse handy, turn off all power by removing the main fuse block, or turning off the main power breaker.
4. Reset the breaker or insert the new fuse.
5. Turn on main power again.

If the breaker trips or the replaced fuse blows immediately, there is probably a defect in the house wiring. Once again, turn off main power. Leave the breaker in the OFF position or remove the blown fuse and leave that socket empty. Turn on main power to the remaining circuits. Troubleshoot the defective circuit as described on page 44.

But if the breaker stays on or the new fuse does not blow, one of the devices that you turned off or unplugged was the cause of the overload. Turn on whatever lights are on the circuit one-by-one. If the breaker trips or the fuse blows, the last light turned on was faulty. Turn off all power to the circuit and troubleshoot as described on page 34. If lights are OK, troubleshoot other appliances on that circuit.

How to map your circuits

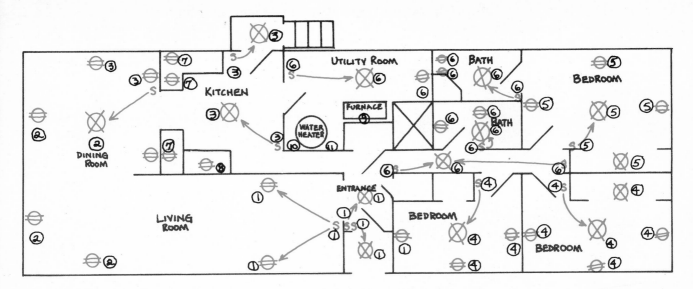

Each circuit breaker or fuse on your service panel should be identified in some way to show the general part of the house it protects. This can be done by putting adhesive-backed paper or plastic stickers next to each circuit breaker or fuse. Or you can make a sketch of the panel layout with circuits identified, and attach it to or keep it near the service panel. In addition to this, however, a map of your house showing exactly which lights, outlets, appliances, etc., are on each circuit can be a great help in planning electrical work and in tracking down troubles. If yours is a new home, the builder or electrical contractor may be able to supply you with a copy of his wiring diagram. The symbols shown above will help you to understand the diagram and enable you to mark it up so that it is more useful to you. If you cannot get a circuit map ready-made, you can make your own. The map you make will probably provide a few surprises. Circuits are often split between rooms—and outlets within a room are split between circuits—in ways you could never have guessed.

To start with, draw a floor plan of your house. Make it floor-by-floor or room-by-room, whichever is more convenient, but be sure to include every area that has electric service. Include porches, garages, outbuildings, etc. Use some system of symbols (the ones on the example are commonly used) to identify every fixture, wall outlet, and switch. Don't forget outside outlets, entrance lights, outdoor floodlights, etc.

Then number each circuit breaker or fuse on your service panel. Next, turn on all the ceiling and wall fixtures and lamps in your house. It is not necessary to turn on major appliances at this time. The procedure is to turn off each circuit breaker or remove each fuse individually. Then determine which lights are off. Mark the number of the circuit breaker or fuse just turned off or removed next to the fixture and switch symbol on your diagram. Next check all the wall outlets in the rooms in which lights went out. Plug a small lamp or work light in each outlet. If the lamp does not light, mark the number of the circuit breaker or fuse next to the outlet symbol on your diagram.

Repeat the procedure for each circuit breaker or fuse on your panel. When you finish, every symbol on your floor plan should have a circuit number next to it. If any symbol has been missed, recheck the area by turning on the light or plugging your work light into the outlet. Next, turn off, one at a time, each circuit that your diagram shows on nearby fixtures and outlets until you find the one that applies.

You will find that some of the circuit breakers you have turned off or fuses you have removed have had no effect on lights or outlets. These are circuit breakers or fuses that protect large appliance circuits. Turn off these remaining circuit breakers or remove these fuses one at a time and check your appliances to find out which one does not work. Remember that furnaces and air conditioners may appear to be off because of temperature-control settings.

Be sure to note on your diagram any circuit uses that are not covered by symbols and may be useful to you. For example, two circuit breakers or fuses are used in electric range circuits. Note on your diagrams which device protects the oven and which the surface burners. When you finish your diagram, store it near your service panel in a protective envelope.

Play it by ear
I ran upstairs and back down to the cellar until my tongue hung out trying to find out which lights and wall plugs went dead when I pulled out each of the fuses in turn. You can save some sweat by using a small plug-in radio (instead of a work light) to check wall outlets. Plug the radio into the outlet you want to check. Tune to a strong station and turn the volume up. Back at the service panel, turn off the circuit breakers or remove fuses one-by-one. When the sound goes off, you have found the circuit.

Practical Pete

2. TECHNIQUES

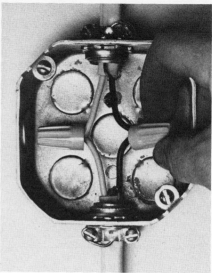

The next few pages show you how to work with the kinds of cables and wire commonly used in home electrical systems. Included are basic techniques and routine tasks that are necessary for any wiring job. These tasks are really easy to do, but practice always helps. Before actually working with a type of wire or cable that is new to you, cut off a short piece and try stripping, joining, etc. Experiment a bit to find out which of the tools you have available are easiest for you to use and which do the best job. A little time spent in trial and error will make the job go faster.

Sheathed cable

Permanent indoor installations are made by running lengths of wire between outlets and switches along or inside walls, floors, and ceilings. An electrical circuit always needs a hot and a neutral conductor plus a ground for safety. When these individually insulated wires (black for hot, white or gray for neutral) are held together inside plastic or metal sheathing, the unit is called cable.

The most commonly used cable for indoor wiring is the flat, white plastic type sometimes called by one trade name, Romex. There are three kinds. Code letters are printed on the plastic. Type NM is designed for normal indoor use; type NMC is insulated well enough to use above ground or in damp areas indoors; type UF can be buried outdoors.

Another kind of cable, usually called BX, is wrapped with a spiral of flexible, galvanized steel armor. This is more expensive and cannot be used outdoors.

Removing plastic sheathing

Place cable on a solid flat surface. Use a utility knife to cut the sheathing along the flat side. Try to make the cut straight and as nearly as possible in the center of the sheathing. The cut should be six to eight inches long. Use enough pressure to penetrate the sheathing but not so much that you cut the insulation on the inner wires.

Peel back the plastic sheathing to the beginning of the cut. Trim off the plastic with wire cutters or large shears. Remove and cut off the paper that is wrapped around the inner conductors. If you accidentally nick the conductors or the ground wire, simply trim off the section and try again.

Removing steel armor

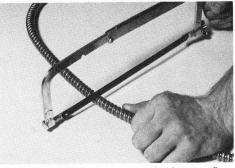

Hacksaw method. If you use a hacksaw, a fine-toothed blade will work best. Cut diagonally across one of the metal ribs. Cut carefully and stop as soon as you have cut through the metal to avoid cutting into the wire insulation.

Next grasp the cable on each side of the cut. Bend the cable back and forth until the armor snaps. Slide the armor off the cable. Unwrap the paper from the inner conductors and cut it away.

Cutting-tool method. First, bend the cable sharply until the armor buckles.

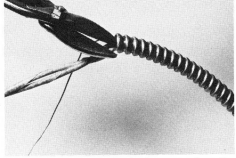

Twist the cable in the direction that will unwind the armor spiral. This causes a section of the armor to spring out at the point of the bend.

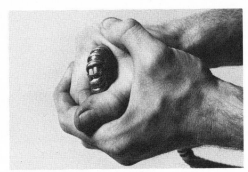

Slip the cutting tool through the armor where it has buckled. Trim away sharp edges. Slide off the end of the armor.

Use the shaping grip in the jaws to reform the buckled cable-end after stripping the paper from the inner conductors and trimming it off.

Trim away or bend the sharp edges of the armor at the point of the cut. Make sure no edges are in position to cut into the insulation in the inner conductors. This is important.

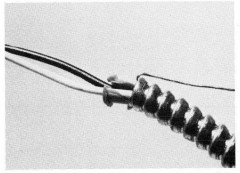

To eliminate the possibility of sharp edges of the armor cutting into the conductor insulation, a fiber bushing should be inserted under the armor, at the point where the conductors emerge.

Choosing the correct wire

Wire load guide

The current capacities listed below may be used as a guide in determining the wire size needed for a job.

Wire size	Amperes	
	Copper	Aluminum*
18	7	—
16	10	—
14	15	—
12	20	15
10	30	25

*Applies to copper-clad aluminum as well.

The wire to use for a particular job will often be specified by your local electrical code. A good source of information on this is the store where you buy your electrical supplies. They know local codes and will be glad to advise.

The table below gives the physical and electrical characteristics of the most common types of wire. Many of them you will already be familiar with.

The wire codes listed in the table below are National Electrical Code designations.

These wire designations are essential to the electrical industry but can be confusing to the "do-it-yourselfer." For example, the wire codes in the table apply to the complete wire or cable as described. But, in addition to these wire codes, there are codes for the individual conductors in the cable. The simplest solution is to check the table to get an idea of what is available. If you have any question about the particular job you are planning, check with your local building department or electrical supplier.

Type UF or NMC

Type AC

Type NM

Type SO

Type HPD

Type SPT

Common wire sizes and uses

Code designation	Common name	Conductor size	Number of conductors	Outer covering	Use (See load guide at left.)
UF	—	12	3 plus ground		
UF	—	12	2 plus ground	Heavy plastic	Underground*
UF	—	14	3 plus ground		
UF	—	14	2 plus ground		
NMC	—	12	3 plus ground		
NMC	—	12	2 plus ground	Waterproof plastic	Damp indoor location
NMC	—	14	3 plus ground		
NMC	—	14	2 plus ground		
AC	BX	12	3 plus ground		
AC	BX	12	2 plus ground	Steel armor	Dry indoor location
AC	BX	14	3 plus ground		
AC	BX	14	2 plus ground		
NM	Romex	12	3 plus ground		
NM	Romex	12	2 plus ground	Nonmetallic reinforced fabric	Dry indoor location
NM	Romex	14	3 plus ground		
NM	Romex	14	2 plus ground		
SO	Heavy-duty cord	16	2 plus ground	Rubber or plastic	Power tools, mowers, etc. Oil resistant
HPD	Heater cord	16/18	2	Plastic, asbestos, fabric	Heaters, toasters, irons
SPT	Lamp cord	16/18	2	Rubber or plastic	Lamps, small appliances, extension cords, etc.

The uses shown in this table apply only to copper wire. Special precautions must be taken if you work with aluminum wire. See page 23 for details.

*Must be protected at the source by circuit breakers or fuses.

Aluminum wire

Aluminum is a less satisfactory conductor than copper. Copper-clad aluminum (aluminum wire with a thin coating of copper) is a reasonable compromise, but copper still has the edge. If your house has copper or copper-clad aluminum wiring, stick with it. Safely adding aluminum wiring to a copper wire system is a tricky business. Avoid it.

If aluminum wire is used in your present electrical system, and the correct switches, receptacles, etc., are used, the system is perfectly safe and efficient. Changes and additions to the system can be safely made using aluminum wire.

The *only* switches and receptacles that are safe to use with aluminum wire are those marked CO/ALR or CU/AL by the manufacturer. Both markings mean that the de-

vice is designed for use with either copper or aluminum wire. The CO/ALR marking is used on devices rated up to 20 amperes. The CU/AL marking is used on devices rated at more than 20 amperes. Aluminum wire must never be used with any device having push-in type wire connections. Push-in type connections can be used only with copper or copper-clad aluminum.

If you use aluminum wire, remember that most recommended wire sizes on appliances or in manufacturers' data are based on copper wire. Unless other data is given for aluminum wire, a good rule is to use aluminum wire two sizes larger than the size specified for copper. For example, if number 14 copper wire is recommended, use number 12 aluminum wire.

Aluminum wiring is potentially dangerous. Check and tighten all connections on switches and outlets periodically. Make sure all switches and receptacles are marked CO-ALR or CU-AL. Replace any that are not.

Removing insulation

Before you can make a connection to a switch or fixture or before you can join two wires, you must remove the conductor insulation or sheathing.

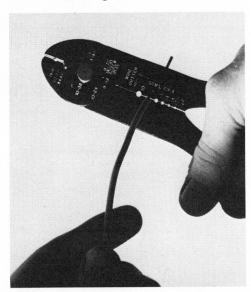

Using a wire stripper is the best method of removing insulation. Place the wire in the proper size hole on the stripper. Close the stripper jaws to cut through the insulation. Rotate the stripper back and forth a quarter turn or so until you can pull the insulation off the wire.

If you are not sure of the wire size, cut off a bit and make a few test cuts. If you can't spare any cable for testing, try one of the larger stripper openings first. Remember, the smaller the wire number, the larger the wire diameter. If you can't pull the insulation free after rotating the stripper, try the next smaller hole, until you get a clean strip and a bare, undamaged wire end.

If a wire stripper isn't available, a knife can be used to remove insulation. To avoid nicking the conductor, cut through the insulation at an angle. Check the conductor closely after removing insulation to make certain the knife has not gouged the metal. If the conductor is nicked even slightly, it will almost surely break before the job is finished. Best to start over now.

Connecting wires to screw type terminals

Terminal connection | Continuous connection

Conductor wire should be formed into a hook for connection to screw type terminals on switches and receptacles. Long nosed pliers are best for the job. The hook should be placed on the screw so that tightening the screw tends to close the hook. The screw turns clockwise, toward the opening in the hook. This type of connection keeps the conductor securely under the screw head. Remember—only black wires (or wires coded black) should be connected to brass terminals.

Don't bet on long shots
It got real cold last winter in the garage, where my workbench is, so I brought an electric heater out there and plugged it into the house wiring via a long extension cord. When I turned it on, the lights dimmed and the heater gave off such a feeble glow that I thought it was broken. I didn't realize that wire itself puts up resistance, and thin wire in a long extension cord drains a lot of power from the system on its way to the appliance. It makes a big difference with an electric heater that uses a lot of watts, but even my power tools won't spin up to their full rpms if the extension cord is very long. Guess I better read chapter 4 and put an outlet out there.

Practical Pete

Joining wires in electrical boxes

For permanent home wiring to switches, outlets or fixtures, wires should be joined only in a wiring box. Three representative types of metal boxes are shown below. There are many variations. Plastic boxes are also available. Your dealer can tell you if your local code allows them to be used in your area.

The general procedure for joining wires in a box is to remove two or more knockout holes in the box, mount the box securely at a point where you wish to make the joint, insert the cables through the opening in the box, secure them with cable clamps, make the electrical connections with solderless connectors, press the wires back into the box, secure a cover on the box. Boxes are available in a wide variety of sizes and shapes. The box to use in a particular location depends on the number of cables to be joined, the kind of mounting available, and whether the box must also provide a mounting for a switch, outlet, or fixture. Check the table below each illustration to determine the box best suited to your installation. As a general rule, unless space is limited, select a box larger than the minimum size. It will be easier to work in and can accommodate additional wiring, if need be, at a later date.

Basic types of boxes

Ceiling box

Typical size (inches)	Max. Number of Wires	
	No. 14	No. 12
4 x 1¼	6	5
4 x 1½	7	6
4 x 2⅛	10	9

Junction box

Typical size (inches)	Max. Number of Wires	
	No. 14	No. 12
4 x 1¼	9	8
4 x 2⅛	15	13

Rules for counting the number of conductors in a box

1. The count refers to each individual conductor, not each cable.
2. Do not count ground wires or jumper wires.
3. Do not count wires from a fixture into the box.
4. *Internal* cable clamps or fixture mounting studs count as one conductor no matter how many clamps or studs there are. External cable clamps do not affect the count.

Wall box

Typical size (inches)	Max. Number of Wires	
	No. 14	No. 12
3 x 2 x 2½	6	5
3 x 2 x 2¾	7	6
3 x 2 x 3½	9	8

Ganging. Wall boxes are designed to be combined into two or more units when more conductor space is needed or when several switches or receptacles must be mounted in one location. Loosen the screw in the flange at the bottom of one box and remove the left side. Loosen the screw in the flange at the top of another box to remove the right side. The two open sides can now be joined by mating the notches with the flanges and tightening the screws at the top and bottom.

Inserting and securing cable

All boxes have knockout holes through which the cables enter the box. Before mounting the box, decide which knockout holes you want to use.

Cables entering the box must be secured with cable clamps. Some boxes have internal cable clamps that will hold either armored or plastic cable. Others don't come with clamps installed, so you must buy and attach separate two-piece clamps.

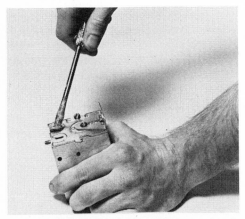

Some knockouts have slots in which you can insert a screwdriver and twist out the knockout.

Other knockouts are solid and must be removed with a hammer and punch.

Preinstalled clamps with steel armor cable. Slide the cable through the knockout hole and under the clamp. Feed the conductors through the upper ring. Tighten the screw in the center of the clamp to secure the clamp against the steel armor.

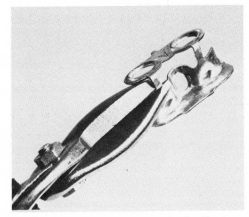

Preinstalled clamps with plastic sheathed cable. Bend and break off the upper ring section. Screw it loosely inside the box next to the knockout hole. Insert the cable through the knockout and secure it by tightening the clamp screw.

Securing clamps to boxes

1. First remove the nut with the pointed "ears" from the threaded end of the clamp. Slide the clamp on the cable and secure to the end of the sheathed portion of the cable by tightening the screw at the side of the clamp. With two-wire cable wrap a separate ground wire around the screw; then tighten.

2. Insert the conductors through the knockout opening and draw the conductors through until the threaded end of the clamp projects into the box. Slip the clamp nut over the conductors so that the ears on the nut will press against the wall of the box when the nut is tightened.

3. After tightening the nut as much as possible by hand, place a screwdriver or punch against one of the ears and tap the screwdriver or punch with a hammer to tighten the nut. Make certain the ears make solid contact with the wall of the box.

Mounting electrical boxes

Wall box, direct mounting.
Metal boxes for outlets or switches can be nailed beside studs through holes in the box or flange on the side. All such boxes must jut out from the stud by the thickness of the finish wall or ceiling material to be applied later, so the box ends up flush with the wall.

Wall box, flange mounted. Some wall boxes have flanges attached to them that can be nailed to the front of the stud. The flange is offset so the face of the box will be flush with the finish wall. Armored ("BX") cable can be looped through the hole in the flange.

Junction box, direct mounting.
Junction boxes that can be nailed or screwed to the sides of studs have knockout holes on three sides and the back so cables can be fed into them from all sides.

Between-stud mounting. When a box must be positioned between two studs, joists or beams (as frequently happens in ceilings) adjustable straps or hangers can be attached to the back of the box and fastened to the structural wooden supports at both ends.

Mounting in new construction is simple and straightforward. Studs and joists are readily accessible and a wide variety of mounting brackets are available. On new construction, wall and ceiling boxes should be installed so that the open front edge of the box will be flush with the finished surface. This means that the front of the box must project into the room beyond the front edge of the stud by an amount equal to the thickness of the wallboard, paneling or plaster that will be fastened to the bare studs as the finished interior surface.

If you can locate a stud in the wall, outlet boxes can be mounted with wood screws. Cut the wall opening even with one side of the stud. Hold the box in position in the opening. Make a pencil mark on the stud to correspond with two mounting holes in the box. Pick one hole near the top of the box and one near the bottom. Remove the box. Use a power or hand drill with a long bit to drill two pilot holes at an

Mounting in existing walls and ceilings requires some ingenuity to keep damage to walls or ceilings to a minimum. Different types of wall boxes are available, some designed to be attached directly to a stud and some that attach to the finished wall material between studs. Either way, position the new box at the same height as other outlets or switches in the room. Make a cardboard template the same shape as the open face of the box. Trace it onto the wall and drill holes in the wall at diagonal corners of the tracing to start the saw cuts.

angle into the stud. The drill bit should be of smaller diameter than the screws you plan to use. About one-half inch depth should be enough. After cables have been installed in the box, reposition it in the opening. Secure it with wood screws through the box mounting holes and into the pilot holes in the stud. Ceiling boxes with mounting flanges can be similarly mounted on joists.

For appearance or convenience, electrical boxes must often be mounted between studs and joists. The devices shown above make the task fairly simple. Expansion brackets located on the box spread when the screw is tightened. When the box is ready for final installation, push the box into the wall so that the ears at the top and bottom hold the front edge flush with the wall surface. If necessary, adjust the ears. Secure the box by tightening the screws on both sides. Tightening the screws spreads the bracket and holds the box in place.

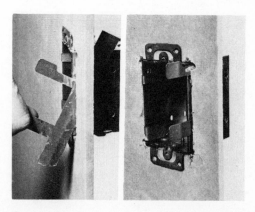

Push the box into the wall so that the ears at the top and bottom hold the box flush with the wall surface. If necessary, adjust the ears. Slide a bracket between the edge of the box and the wall opening. Hold the bracket by the arms and pull it tight against the wall. Bend the arms toward the inside of the box. Install another bracket on the other side of the box in the same way. Use pliers to bend the arms tight and flat against the sides of the box.

Joining wires together

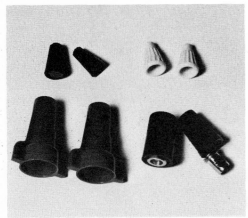

Solderless connectors

Solderless connectors are used to make electrical connections between the conductors within the metal box. Plastic screw-on connectors, sometimes called "wire nuts," are the most common type and are satisfactory for most uses. They come in a variety of sizes to accommodate different combinations of wire thicknesses and number of conductors to be joined. They are so inexpensive you can keep a variety of sizes on hand to be sure you have one that will make a good electrical contact and not work loose. (If there is any vibration, wrap tape around the wires and the base of the nut. If you ever need to rewire, unscrew the nut and screw it back on the new wires.)

Plastic screw-on connectors contain a tapered metal insert with spiral grooves that grip the ends of the wires when you screw on the connector. The bare conductors do not have to be twisted together first, but they must be trimmed in length so that when the connector is screwed down tight over them, no bare wire is exposed.

Hold the conductors together and approximately parallel. Twist the connector onto the wires clockwise by hand. Check each conductor to make certain it cannot be pulled free.

Ceramic screw-on connectors are used in high temperature locations, such as heaters and irons.

Metal set-screw type. Wires are inserted under the set screw. The set screw is tightened and the plastic cap is screwed on the connector afterwards.

Heavier connectors for larger size wire are tightened with a wrench to clamp the wires together. These connectors must be insulated by wrapping with tape.

Wiring lamp, appliance plugs

Use the Underwriters' knot to attach lamp cord to plugs with screw terminals.

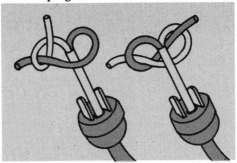

1. Feed the end of the cord through the plug. Separate the conductors for about three inches. Tie the knot as shown.

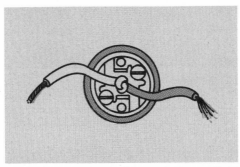

2. Strip insulation from each conductor. Twist stranded wire tightly.

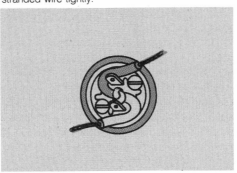

3. Pull the knot securely back into the plug. Route each conductor around a prong.

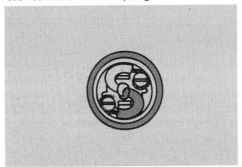

4. Loosen the screw terminals and wrap one bare conductor around each. Be sure the direction you wrap the conductor around the screw is the same direction you turn the screw to tighten it securely.

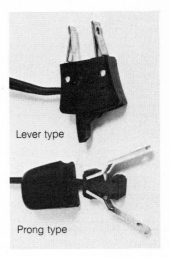

Lever type

Prong type

Crimping-type plugs for flat wire
Several kinds of plugs are available that accept flat wire and make connections automatically without your having to take them apart and strip the insulation off the ends of the wires. Most of them grip the wires when you either close a lever or squeeze two prongs together.

To use these plugs, cut the lamp cord squarely at the end. Some designs also require that you make a short slit between the two insulated wires to divide them. Do *not* remove insulation. Insert the wire into the opening on the plug as far as it will go. Fully depress the lever on the plug, or squeeze the prongs together and insert them into the housing of the plug. The lever or prongs make the electrical connection in the plug by forcing sharp metal points through the insulation to make contact with the conductors. The lever or prong action also crimps the wire to provide some mechanical strength.

If the cord is likely to be plugged and unplugged frequently, a screw type plug with a molded grip, wired with the Underwriters' knot, is stronger and will last longer than automatic crimping plugs.

Splicing wires

Cabling to wall outlets, switches, and large appliances should never be spliced. All connections should be made within metal or plastic boxes, as previously described.

Occasionally, it may be necessary to splice wires when repairing or modifying appliances or for emergency repairs. The proper way of making safe, secure splices is to make a good mechanical and electrical joint—that is, strong enough not to pull apart and tight enough so there is no loss of voltage. As this is tricky work, a continuous run of new wire is always better than any splice.

Taping wire splices

Spliced and soldered joints should always be taped. The right amount of tape to use on a joint is the amount that will provide insulation about as thick as the original insulation on the wire. A good brand of plastic electrical tape is best for wire joints.

Apply the tape by wrapping it diagonally along the joint starting on the insulation at one end.

Plastic tape sticks best if it is kept taut while wrapping. Continue the tape for an inch or so on the insulation at the other end.

Make as many wraps as necessary to build the tape to the proper thickness. Cut or tear the tape at the end of the last wrap and press it smooth around itself.

The twist splice

For light wire, when two wires are joined, cross about two inches of each end of prepared wire. Bend the ends of the wires over each other at right angles and twist them around each other.

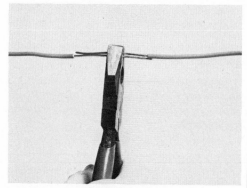

For heavy-gauge wire, two pairs of pliers are needed to make sure the connection is tight. Use one pair of pliers to hold the wires at the beginning of the twist.

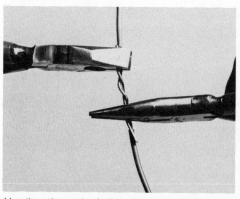

Use the other pair of pliers to twist the wires. Use wire cutters to trim off the excess wire so that no sharp ends can penetrate the tape. Solder the wires at the twist and tape them.

The pigtail splice

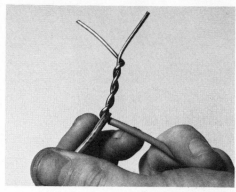

1. Strip off at least 1½ inches of insulation from the end of each wire. Twist the wires together tightly starting at or near the first bit of exposed wire.

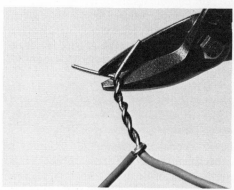

2. Trim off sharp points protruding from the end of the twist. Solder the twisted wires at the point where the twist began.

3. Bend them parallel to one of the conductors and tape the bare splice from the end of the insulation on one side to the beginning of the insulation on the other side.

Splicing three or more wires

The pigtail type of splice is best when joining three or more wires. The thing to guard against when more than two wires are involved in the twist is the tendency for one or more of the wires to remain fairly straight while the others are wrapped around it. When this happens the straight conductors can be pulled free of the splice fairly readily. The way to prevent this is to make certain the twist is started with all the wires bent at approximately a right angle. Then if the bent wires are interlocked and held with pliers, the twist will continue as started. Solder and tape the splice.

Starting a three-wire pigtail splice. To interlock all three wires bend each one at a right angle when you make the first twist. A straight wire will pull out under relatively little stress.

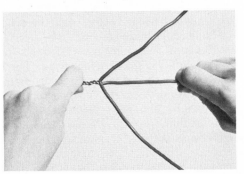

Finishing a three-wire pigtail splice. Before soldering and taping the exposed wire, pull on each of the three wires to be sure they are firmly interlocked.

ap splice

Sometimes it is desirable to join one wire to another at some midpoint without cutting the second conductor. The tap splice can be used in this case. Remove insulation from about two inches of the continuous run wire. Wrap the joining wire tightly around the continuous wire. Solder and tape.

If the joining wire is stranded, the strands may be separated into two bunches and then wrapped in opposite directions along the continuous wire.

Don't get caught in a lousy joint
Personally I think wire nuts are just as good as splicing, soldering, and taping—and a heck of a lot less work. But my granddaddy said, "Solder," so I soldered. Trouble was, I touched the hot iron to the solder and tried to drip it onto the wire. Even when the drop didn't miss altogether, it hardened so fast on the cold copper that it didn't run into the little crevices between the twisted wires, and they pulled apart. "Heat the *wires*, sonny boy," he said, "and hold the solder on *them* so it melts into them." It works fine that way, but it's a little hard to take that kind of talk when you're going on 37 years old.

Practical Pete

Soldering

For a good solder joint in house wiring you need a high heat soldering tool and rosin core solder. The solder should be marked as suitable for electrical use. A high heat electric soldering gun or iron can be used, but pencil-flame propane soldering torches work faster. Of course considerable care must be taken with the open flame to avoid injury or damage from fire. The trick with house wire is to heat the joint as rapidly as possible.

The large diameter wires are good heat conductors, as well as electrical conductors. They tend to carry heat away from the joint. This causes scorched insulation and can cause a poor solder joint.

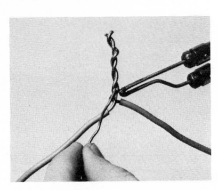

Hold the wires to be soldered tightly against the tip of the iron. Unwind a few inches of solder from the spool and hold the end of the solder against the wire, not against the iron. To make a good joint the wire must become hot enough to melt the solder on contact.

Inspect the solder closely in a good light after soldering. The solder should look smooth and a bit shiny. If the solder looks dull and "grainy," the chances are a good joint has not been made. Reheat the joint until the solder flows again. Recheck the joint for strength after it has cooled.

The soldering procedure using a propane torch is essentially the same. Make certain the flame is adjusted and positioned for electrical work in accordance with the manufacturer's instructions; its intense heat can melt the wires. Use a soldering tip that pinpoints the flame.

Conduit

Conduit is often used to hold and protect house wiring. In some localities conduit is required by the local code. Conduit protects house wiring from damage better than steel armor or plastic sheathing. However, it is more difficult to install.

There are three types of conduit available: thin-wall metal conduit, rigid threaded conduit, and plastic conduit.

The most common type of conduit for house wiring is the thin-wall type. Thin-wall conduit is too thin for threaded joints. It is joined to other lengths of conduit and to boxes by pressure-type fittings. Thin-wall conduit is sold in ten-foot lengths in either one-half inch or three-quarter inch (outside) diameter. The one-half inch conduit can contain four No. 14 wires or three No. 12 wires. Three-quarter inch conduit accommodates four No. 10 or five No. 12

wires. These capacities are for individual wires, not pairs. The wires used in conduit are the same as the individual conductors found in steel armor cable and plastic sheathed cable. Wires in conduit must follow standard coding. In a two-wire conduit circuit you need one black wire, one white wire, and one ground wire. A three-wire circuit requires one black wire, one white wire, one red wire, and one ground wire.

The general procedure for using thin-wall conduit is similar to the use of steel armor cable. The big difference is that conduit cannot be "snaked" through openings in ceilings and walls. You must have full access to joists and studs to install conduit. So you probably won't want to use it unless your local code requires it.

The tools you need and the procedures for using conduit are shown on these pages.

Cutting

Thin-wall conduit can be cut with a special cutter. To use the cutter, clamp it around the conduit. To cut the conduit, tighten the knurled nut in the handle. As you tighten, force the cutter around the conduit. The sharp cutting blade cuts a groove that deepens with each revolution, making a smooth, quick cut. After cutting through the conduit, file off any burrs around the edge of the cut.

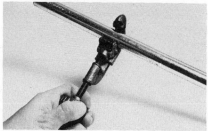

1. Turn knurled nut to open jaws of cutting tool and tighten them again onto conduit.

2. Rotate cutter around conduit as you tighten until blade scores through.

Bending

Thin-wall conduit can be readily bent by using a special tool designed to make a smooth, even bend with little effort. The more bends in a run of

conduit, the more difficult it is to "fish" the wires through. Plan the conduit run carefully to avoid sharp bends and to make as few bends as

possible. Never have more than four right angle bends between openings. Follow the manufacturer's instructions for the type of bender you use.

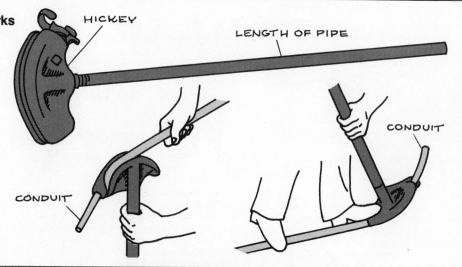

How a conduit bender works

Screw a 30-inch length of threaded pipe into the bender head (sometimes called an "electrician's hickey")

Insert the conduit into the bender through the hook at the top of the head. The hook marks where the bend will start.

Put one foot on the conduit near the head and lever the pipe handle backward, checking the angle of the bend as you go.

HICKEY

LENGTH OF PIPE

CONDUIT

CONDUIT

Supporting and coupling

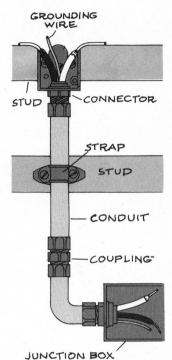

Typical conduit installation

Support lengths of conduit with a pipe strap every 6 feet for inside runs; 10 feet on exposed runs. Mount empty conduit in place and make all couplings and connections to boxes *before* threading the insulated indoor wire through it.

Join two lengths of unthreaded conduit with couplings. One type has three nuts in a row. Force the ends of the thin-wall conduit into each end of the coupling and tighten the center hex nut. The other type has two set screws.

Connecting to boxes

To attach conduit to a junction box, use a ring-nut connector. Insert the conduit in the connector and tighten the hex nut on the connector to squeeze a compression ring inside the nut. This secures the connector on the conduit.

Next, insert the connector through the knockout on the box. Thread the ring nut on the end of the connector projecting into the box. Tighten the ring nut by placing a screwdriver against one of the "ears" on the ring nut. Tap the screwdriver with a small hammer to tighten the ring nut.

Threading wires through conduit

A fish tape is a thin, flexible metal tape with a hook on one end. The tape is usually packaged on some type of reel. The tape is used to pull wires through conduit or through openings in walls. For conduit use, the tape is inserted in one conduit opening and worked through to the next opening. The wires to be drawn through the conduit are bent around the hook on the fish tape.

If the run is long and has a few bends it is a good idea to wrap some electrical tape around the wires to hold them on the hook.

The tape is then reeled in to draw the wires through the conduit. A slow, steady pull is less likely to kink the tape or jam the wires than is a series of sharp tugs.

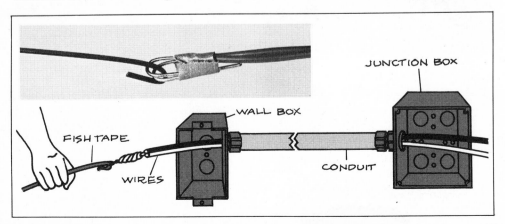

3. SIMPLE REPAIRS

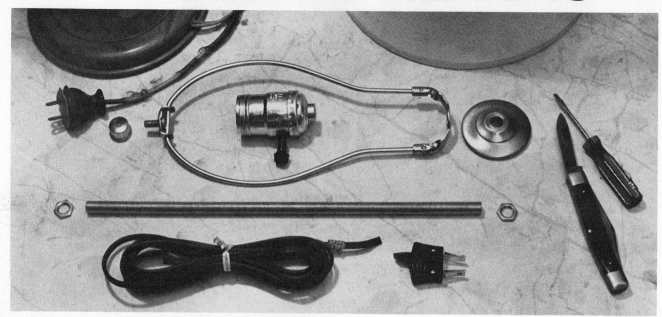

Rewiring a lamp

Perhaps the most common household electrical problem is the table or floor lamp that will not light, flickers on and off, or sometimes works and sometimes does not.

Fortunately this common problem is easily corrected. The possible sources of trouble are the same in all lamps. The first thing to check, of course, is the bulb. Make sure it is fully screwed into the socket. If that doesn't do it, switch bulbs with a lamp that you know is working. If the first lamp now works and the second one doesn't, the bulb has burned out. Or, if you have a continuity tester, you can check the bulb by placing the tip of the probe on the center of the base of the bulb and touching the alligator clip to the threaded part of its base. If the light in the tester goes on, the bulb is good.

If bulb life seems unusually short in a lamp, the lamp may be faulty. (The average life of a 100-watt bulb is 750 hours of operation.) Try the bulb in another lamp or fixture, as suggested above, before discarding it. If the bulb is not at fault, the cause of the trouble is either the lamp socket and switch or the lamp cord. You can check these with a continuity tester (page 34) and replace only the defective item. You can also purchase lamp rewiring kits with various socket-and-switch combinations. The wiring for three-way sockets is the same as for single-wattage sockets.

Single-socket lamp

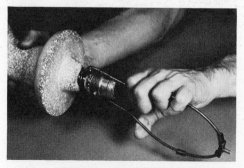

1. Unplug the lamp from the wall outlet. If the lamp has a harp (shade support bracket), remove it by lifting up the ferrules at each side of the base of the harp, compressing the bracket slightly, and lifting it free so you can get at the socket more easily.

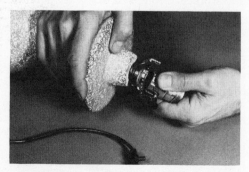

2. Press in the side of the socket near the base (where the word PRESS is sometimes marked). Twist slightly and pull up to remove the metal shell. Then remove the cardboard insulator under the metal shell to expose the socket terminals.

3. Push the line cord in at the base of the lamp and pull up on the socket to inspect the line cord and to gain some slack to work with. Loosen the screw terminals and disconnect the line cord conductors from the socket.

4. Check the insulated portion of the line cord near where it is connected to the socket. If the insulation is brittle and cracks when bent, either the cord is quite old or the socket is getting too hot. In either case, both socket and cord should be replaced. If the cord looks OK and tests OK, it can be reused.

5. The opening in the bottom of the lamp base is often covered with a felt pad. It will be easier to remove the old cord and to thread the new cord through the lamp if you peel off the pad. Peel carefully to avoid tearing the felt. The pad can be glued back on the base when you have finished wiring and testing.

6. To remove the power cord, simply pull the old cord out of the base after disconnecting it from the socket. If the cord does not pull out readily, check for strain knots at the top where the cord enters the base and at the bottom where the cord leaves the base.

7. Remove the old socket base (loosen the set screw on the side if there is one) by unscrewing it from the threaded pipe in the lamp. Feed the new cord through the base, tie a simple strain (overhand) knot in the cord inside the base opening, and then feed the cord up through the threaded pipe.

8. Slip the new socket base over the line cord, thread it onto the pipe, and secure it by tightening the set screw. The set screw should keep the cord from slipping back through when you pull on the plug end. If it slips, separate the two halves of insulated cord and tie an Underwriters' knot (page 27).

9. To separate the two line cord conductors at the socket end of the line, grip each conductor half with your fingers or with pliers and gently separate them along the groove. Make the tear about 2 inches long. Strip about 1 inch of insulation from each conductor. Clean the bare wires by scraping lightly with a knife or razor blade. If the conductor is braided, twist it tightly.

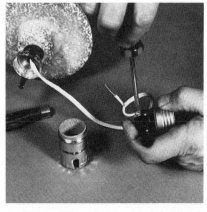

10. Wrap each conductor around one of the screw terminals on the socket. Wrap the wire so the insulation is close to the terminal, and wrap in the direction in which you will turn the screw to tighten it. One turn around the screw is enough. After you have tightened the screw, trim off the excess exposed conductor as close to the terminal as possible. Before assembling the socket, check and retighten each screw terminal.

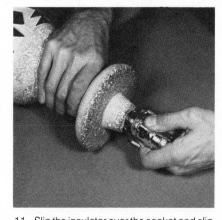

11. Slip the insulator over the socket and slip the outer metal shell over the insulator. Push the excess line cord down through the socket base and snap the metal socket shell onto the base. Replace the felt pad on the base. If the lamp does not operate properly, unplug it immediately and recheck your work, testing if necessary to locate the trouble. Replace the harp. Connect a plug to the line cord (page 27).

Testing the parts of a lamp

What it takes

Approximate time: A few minutes for testing; up to half an hour for replacing switches.

Tools and materials: Continuity tester (page 13) for testing; screwdriver, pliers, knife, and new switch for replacing.

The electrical parts of any lamp can be tested with a continuity tester. In the majority of cases, testing will identify the defective part. You then have to purchase and replace only one item. This makes the job quicker and more economical. Keep in mind, however, that the small flashlight battery in the tester does not duplicate the relatively high voltage in the line and resultant heat of actual operation. If the low-voltage tests do not reveal the source of the problem, check the cord and the plug; then replace all the electrical parts because some may be defective under high voltage.

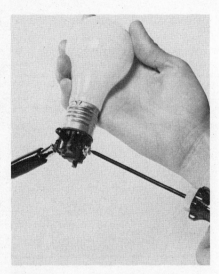

Socket. Put a bulb you know to be good in the socket. Connect the alligator clip to one terminal. Hold the probe against the other terminal. Turn the switch on and off. The tester should light steadily when the switch is on. Jiggle the switch a bit. If the tester flickers, the switch is defective and should be replaced.

Three-way bulb socket. Connect the alligator clip to the brass-colored terminal. Touch the probe to the small tab near the edge inside the socket. The tester should light when the switch is turned to the first ON position. Move the probe to the metal tab in the center of the socket. It should light with the switch in the second ON position. In the third ON position, the tester should light when touched to both tabs in turn.

Plug. Connect the alligator clip to one of the bare conductors. Touch the probe to each of the plug prongs. The tester should light when one—and only one—prong is touched. Hold the tester in place on the prong and jiggle the cord. If the tester flickers, the cord is defective and should be replaced. Connect the alligator clip to the other conductor and repeat the test.

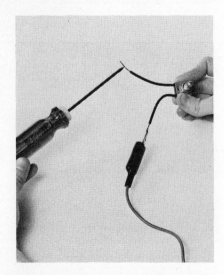

On-off switch. Connect the alligator clip to one of the switch leads. Hold the other switch lead against the probe. Turn the switch off and on several times. The tester should light steadily when the switch is on and should not light up when the switch is off. Replace the switch if the tester flickers in either position or does not light at all.

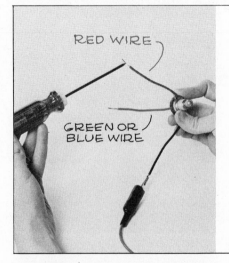

RED WIRE

GREEN OR BLUE WIRE

Switch position	Touch probe to wire listed	Tester light
1	Red	OFF
	Green or Blue	OFF
2	Red	ON
	Green or Blue	OFF
3	Red	OFF
	Green or Blue	ON
4	Red	ON
	Green or Blue	ON

Combination switch. Connect the alligator clip to the black lead. Move the switch through all four positions. At each position, touch the probe to each of the other two leads. If the switch is OK, the tester will light or not light as indicated in the chart above. Although the color of the three leads may vary, the on-off pattern for any three-bulb lamp switch will be the same. If the tester flickers when it should be on, does not light when it should, or lights when it should not, the switch is defective. Repeat the test to be certain the results are correct.

Wiring multiple-socket lamps

Two-socket lamps

There are two basic types of two-socket lamps. In one type, the two sockets are contained in one molded plastic or metal package. In this type the sockets are connected internally, and only two wires—one black, one

white—must be connected to wire both sockets. The switch is separate from the socket. The other type of two-socket lamp has two separate sockets. The sockets may have built-in switches—as in the single-socket lamp—or they may be wired to a common switch.

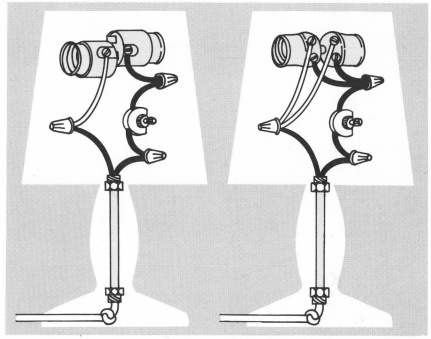

Two-socket lamp with prewired two-socket unit and separate on-off switch (both bulbs on or both bulbs off).

Two-socket lamp with separate sockets and rotary switch (both bulbs on or both bulbs off). If sockets have screw terminals, install black and white jumpers as shown. Connect black jumpers to brass terminals and white jumpers to chrome terminals.

Three-socket lamps

Three-socket lamps—such as pole lamps and swag lamps—usually have switches that allow one, two, or three bulbs to be turned on as desired. This

makes the wiring a bit more complicated, but it can be easily done with the help of the diagram. Sockets in these lamps are prewired with black and white leads.

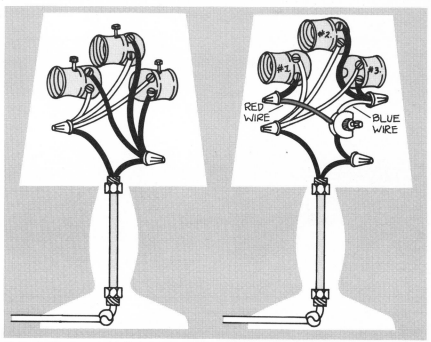

Three-socket lamp with individual on-off switch in each socket. If sockets have screw terminals, install black and white jumpers as shown—black jumpers to brass terminals, white jumpers to chrome terminals.

Three-socket lamp with separate four-position rotary switch. The switch positions are: off, only bulb #1 on, only bulbs #2 and #3 on, all bulbs on. Switch wires are usually black, red, and blue. Black connects to the line, red connects to one socket, blue to the two other sockets.

Replacing a switch

What it takes

Approximate time: Fifteen minutes or more.

Tools and materials: Screwdriver and switch or receptacle with a green grounding terminal.

Wall-mounted toggle switches usually give a warning before they fail completely. When you snap on a light and there is a brief delay before the lamp lights, or if the lamp flickers a bit when turned on, the switch is approaching the end of its useful life.

Switches that are used the most fail first. When a frequently used switch shows signs of failing, it is probable that other much-used switches of the same age will also need replacement soon. It may save time and trouble to purchase and install several switches at once, rather than wait for unexpected failure.

When switches need replacement, take the opportunity to consider changing to other types of switches. See pages 54 through 61 for a rundown of your options.

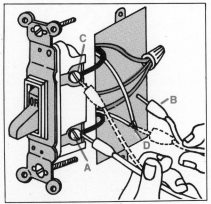

Testing for power in the circuit. Turn off power to the switch at the circuit-breaker or fuse panel. Touch one probe of voltage tester to one of the switch terminals (A). Touch the other probe to the wall box (B). (If the switch has a green-tinted terminal at the bottom, touch the probe to that.) The tester should not light. Touch one probe to the other switch terminal (C) while holding the second probe on the box or green-tinted terminal (B). Again the voltage tester should not light. If there is a white wire in the box, carefully remove the plastic solderless connector joining the white wire. Do not touch the bare wire. Touch one probe to the bare white wire (D) and the other probe to each of the terminals. The tester should not light.

Removing the old switch. Dismantle the switch by removing the wall plate, then the mounting screws at the top and bottom. Pull the switch out of the box. Loosen the screw terminals and remove the electrical connections. If the switch has push-in type connections, insert a small screwdriver in the slot next to the hole where the conductor enters the switch. Press in on the screwdriver and pull the conductor out. Notice carefully which wires connect to which terminals so you can replace them the same way on the new switch. A white wire may be painted black or wrapped with black tape to show it is hot. Plain ON-OFF switches do not need neutral conductors. They break the flow of current through hot conductors only.

Testing the old switch. Use a continuity tester to check whether a switch is good or not. Connect the alligator clip to one of the brass-colored terminals and touch the probe to the other brass-colored terminal. Flip the switch on and off several times and jiggle the toggle. If the switch is good, the tester light will respond clearly to the ON and OFF positions of the toggle without flickering. Move the alligator clip to the switch mounting bracket (or green-tinted grounding terminal if it has one). Touch the probe to each of the brass-colored terminals in turn and flip the switch from off to on several times. If the switch is good, the tester will not light at any time during the test.

Installing the new switch. The replacement switch should have—in addition to the two brass-colored terminals—one green-tinted grounding terminal (a recent requirement of the electrical code for new installations). If the old switch had a green-tinted terminal, simply connect the green wire to the green terminal on the new switch.

Next, connect the black wires (one may be a black-coded wire) to the brass-colored terminals on the switch. Remount the switch in the wall box. (If a grounding jumper wire must be added, see the box at right.) Replace the cover plate, turn on the power, and test the switch.

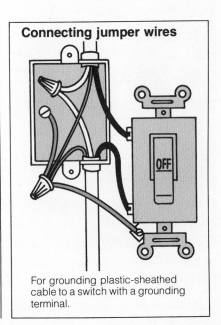

Connecting jumper wires

For grounding plastic-sheathed cable to a switch with a grounding terminal.

Replacing a wall outlet

Unlike switches, wall outlets have no moving parts, so they generally last for many years. They can, however, develop trouble. Plastic parts can become dry and brittle and can chip or crack when plugs are inserted or removed. Metal electrical parts can become loose and make poor contact with plugs.

The present electrical code requires that all wall outlets be the three-prong type with a green-coded grounding screw. Be sure the replacement you purchase is this type. Before touching any part of the outlet, turn off power to the outlet. Then follow the testing procedure shown below.

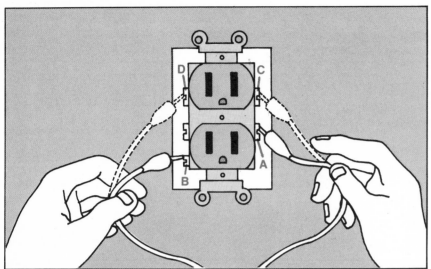

Testing for power in the circuit. Turn off power to the circuit at the main service panel. Take off the faceplate. Touch one probe of a voltage tester to one of the brass-colored terminals (A); touch the other probe to the wall box. If the outlet has a green-tinted terminal at the bottom, touch the probe to that (B). The tester should not light.

Next, touch one probe to the other brass-colored terminal (C) while holding the other probe on the box or green-tinted terminal. Again the voltage tester should not light. Touch one tester probe to one of the chrome-colored terminals (D); touch the other probe to each of the brass-colored terminals, as in A and C. The tester should not light.

Connecting jumper wires

The procedure for adding a grounding jumper to a wall box depends upon the type of cable in the box.

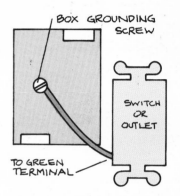

If the power cables to the wall box are the steel-armored type, simply connect a jumper of green-insulated or bare wire to a grounding screw in the wall box. Connect the other end of the jumper to the green-tinted screw on the switch or outlet.

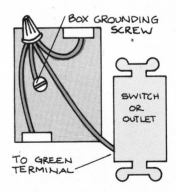

If plastic-sheathed cable is connected to the box, the bare grounding wires from the cables, including any already connected to the box, and two bare or green-insulated jumpers should be joined with solderless connectors. Connect one of the jumpers to a box grounding screw. Connect the other jumper to the green-tinted screw on the switch or outlet.

Removing the old outlet. After taking off the faceplate and testing, remove the mounting screws at the top and bottom. Pull the outlet out of the box. Loosen the screw terminals and undo the electrical connections. If the outlet has push-in type connections, insert a small screwdriver in the slot next to the hole where the conductor enters the outlet. Press on the screwdriver and pull out conductor.

Installing the new outlet. If the outlet you removed had a green-tinted terminal, connect the existing green-insulated wire to the green terminal on the new outlet. Connect the black wire or wires to the brass-colored terminals, and connect the white wire or wires to the chrome-colored terminals. Remount the outlet in the wall box. If a grounding jumper must be added, see box (right).

Replacing small ceiling fixtures

What it takes:

Approximate time: Forty minutes if the new hardware fits the old—more if mechanical adapters are necessary. Allow an extra half hour for chandeliers.

Tools and materials: Voltage tester, screwdriver, locking pliers or wrench, wire stripper or knife, solderless connectors (if you can't reuse the old ones), and a wire coat hanger bent into an S-hook.

When redecorating a room, it is often helpful to temporarily remove fixtures. The first step is to turn off power to the fixture at the circuit-breaker or fuse panel. Do not depend on the wall switch to remove power. Use a voltage tester to check wires before you touch them. The procedure for checking wires in a ceiling box to make sure power is off is basically the same as that used for wall outlets (page 37). Carefully remove all solderless connectors and check between all wires and ground. (The ceiling box should be grounded.) Next, check between all black (or black-coded) wires and all white wires. If the tester lights at any time, voltage is present in the box. Do not continue until the source of the power has been discovered and turned off.

The electrical connections for small ceiling fixtures are quite simple. The box may contain other wiring, but you need only disconnect the two wires that lead to the fixture. To reinstall the fixture, or to replace it with a new one, connect the fixture wires to the same two power wires. If the fixture wires are color coded, connect the black fixture wire to the black power wire. Connect the white fixture wire to the white power wire. Some fixtures have wires of the same color. In this case either fixture wire can be connected to either power wire by using solderless connectors.

Screw-mounted fixtures

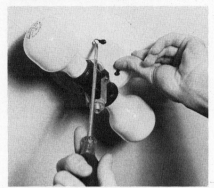

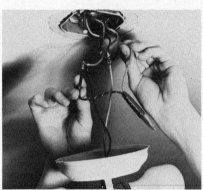

1. This type of fixture is held in place by two long screws inserted through slots in the canopy and threaded into ears on the ceiling box. To remove the fixture, loosen the screws (you need not remove them). Turn the canopy slightly so that the screw heads line up with the larger end of the mounting slot.

2. The screw heads will now pass through the larger opening and the canopy can be lowered. With the glass or plastic globe removed, small fixtures are light enough to hang by the power wires briefly while you make a voltage check to be sure power is off in the ceiling box.

3. When you are sure power is off, support the canopy with one hand. Remove the solderless connectors that attach the fixture wires to the power wires. Check them with a voltage tester. Touch a probe to each. The tester should not light. Untwist the wires and remove the fixture.

Strap adapter

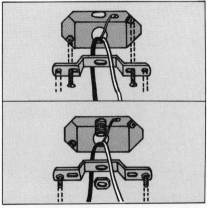

Center-mounted fixtures

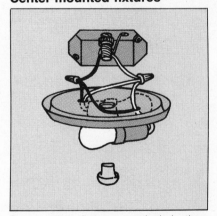

1. The mounting slots in the canopy of a replacement fixture may be too widely spaced for the mounting screws to be threaded into the box ears. In this case, a strap can be mounted on the box. The strap provides threaded holes at each end. Measure the distance between the mounting slots in the canopy, and purchase a strap that matches the spacing.

2. The strap can be attached to the ceiling box in two ways. The strap at top is attached with screws put through the strap slots and into the threaded ears on the box.
3. If the box has a threaded nipple (bottom), the strap can be placed over the nipple and secured with a locknut.

Some ceiling fixtures—particularly those having two sockets—are center mounted. The ceiling box contains a threaded nipple. For mounting, the canopy is placed over the nipple. An end cap is then threaded on the nipple to secure the fixture. To install this type of fixture on a box without a center stud, put a strap-and-nipple combination on the box first.

Replacing chandeliers

Chandeliers differ from small ceiling fixtures in the way they are suspended from the ceiling box and in how they are wired internally. Chandeliers frequently weigh 10 pounds or more, in which case they must be attached to a ceiling box containing a threaded metal stud, or nipple. If you are replacing a small chandelier with a large one, and the ceiling box does not contain a stud, you may be able to add one; otherwise, the box must be replaced with one that can accommodate a stud (pages 62 to 69). **A chandelier weighing less than 10 pounds** can be mounted with a strap and nipple. The nipple is attached to the strap with locknuts. The strap is then attached to the box with screws put through the strap and into the threaded ears on the box, as shown at the bottom of the opposite page. **Mount fixtures weighing more than 10 pounds** by threading a hickey (which resembles a C-shaped bracket with threaded holes at top and bottom) on the box stud and threading a nipple into the hickey. Thread them so that the nipple will project through the canopy enough to allow the collar to be securely threaded onto the nipple.

Chandeliers are fully wired when you purchase them. Only two wires from the chandelier need be connected to the two power wires in the ceiling box. If the chandelier wires are color coded, wire the black chandelier wire to the black power wire. Wire the white chandelier wire to the white power wire. If both chandelier wires are the same color, use either.

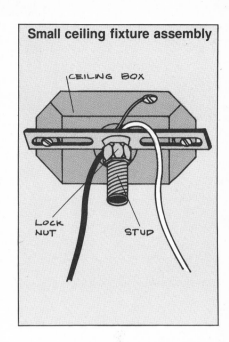

Small ceiling fixture assembly

CEILING BOX

LOCK NUT

STUD

Replacing sockets

Occasionally one or two sockets on a chandelier may become defective. This is sometimes the case when a bulb flickers or does not light. First thing is to check the bulb in another socket or fixture. If the bulb is OK, the problem is in the socket or wiring. **Turn off the power to the chandelier.** Then loosen the collar so that the canopy can be dropped to expose the box wiring. Do not allow the chandelier to hang by the power wires. Support it by making a hook of heavy wire (a wire clothes hanger works well). Attach the hook to the chandelier and the hickey. Take care that the hook does not come in contact with any exposed conductors.

Make certain the power is off (page 38) before you disconnect the wires.

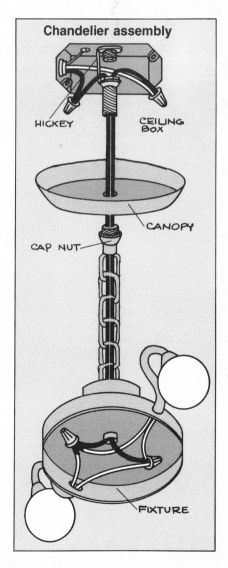

Chandelier assembly

HICKEY

CEILING BOX

CAP NUT

CANOPY

FIXTURE

To replace a screw-terminal socket, loosen the terminal screws and disconnect the wires. Tag or mark the wire that was connected to the brass terminal. Remove the socket by unscrewing it from its threaded mount or by loosening the setscrew in the base. Lift the socket from its mounting base. You can probably obtain an exact replacement socket from your electrical dealer or from a lighting-fixture store. Install the new socket on the mounting base. Connect the wire you tagged or marked to the brass terminal. Connect the other wire to the chrome terminal. Replace the outer shell.

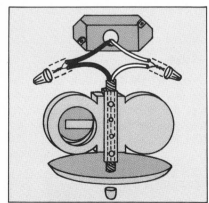

To replace a prewired socket, first disconnect the wires in the center section of the chandelier in order to remove the socket. Remove the lower cap nut and canopy to expose the socket wiring. Then take off the solderless connector joining the socket wires to the other fixture wires. If the wires aren't color coded, tag or mark the wire connected to the black socket wire. Free the socket wires. Unscrew the socket from its threaded mount or loosen the setscrew in the base. Remove the socket and its wires. Install the new socket by routing the socket wires to the center section. Mount the socket. Connect the socket wires to the other fixture wires.

Troubleshooting fluorescents

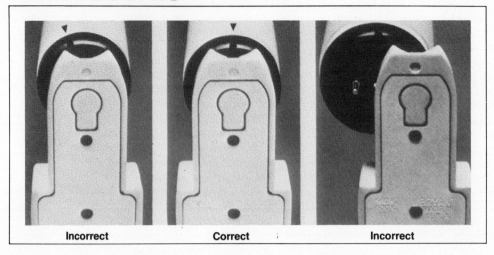

Incorrect   Correct   Incorrect

The bee in the ballast
We decided to put fluorescents in the family room to save electricity. Our teenager's super-sophisticated stereo equipment is set up there too. The lights looked great, but I didn't think to find out the noise level of the ballast beforehand. The buzz was so bad, at first our teenager thought his amp or speakers had blown. Guess which had to go?

Practical Pete

Correct installation

The most common cause of fluorescent lamp trouble is incorrect installation of the lamps in the holders. Fluorescent lamps have marks on each end to indicate proper positioning in the fixture.

The lamp on the left in the photographs above was inserted into the holder correctly but was not rotated enough. The center lamp was correctly installed. The lamp on the right is hanging by one pin.

Noise level

All fluorescent ballasts generate a hum. It can range from barely audible to an annoyingly high level. Ballasts have a sound rating ranging from A (low) to F (high). If the hum level is an important consideration, ask about the sound rating when buying the fixture. The chart, right, recommends rating levels for some living and working areas.

Application	Sound rating
Residence, TV-broadcast studio, church	A
Library, classroom	B
Commercial buildings, stockrooms	C
Retail stores, noisy offices	D
Light industry, outdoor lighting	E
Street lighting, factory	F

Discoloration

Various patterns of light and dark gray appear in the ends of most fluorescent tubes. Although discoloration generally increases with age, it does not affect the operation of the lamp. You can ignore it if the lamp is concealed by a plastic cover or is located behind a cornice or valance. But if the lamp is exposed and discoloration is objectionable, the lamp ends should be concealed in some way or the tube replaced.

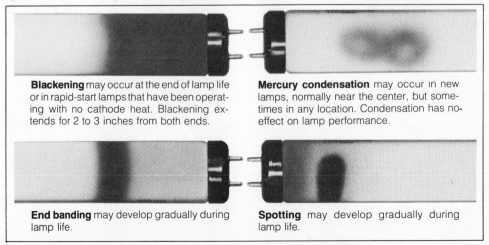

Blackening may occur at the end of lamp life or in rapid-start lamps that have been operating with no cathode heat. Blackening extends for 2 to 3 inches from both ends.

Mercury condensation may occur in new lamps, normally near the center, but sometimes in any location. Condensation has no effect on lamp performance.

End banding may develop gradually during lamp life.

Spotting may develop gradually during lamp life.

Troubleshooting checklist for fluorescents

The troubleshooting chart below lists some of the more common problems that develop in fluorescent lighting and tells how they may be corrected. Unless otherwise indicated, the cause and correction apply to all types of residential fixtures.

Problem	Cause	Correction
Lamp will not light.	Fuse blown or circuit breaker tripped.	Replace fuse or reset circuit breaker.
	Lamp defective or burned out.	Replace lamp.
	Lamp not properly installed in holders.	Check and install lamp correctly (see page 40).
	Starter defective or wrong size.	Replace starter.
	If fixture is on a dimmer circuit, dimmer is defective or improperly adjusted.	Replace or adjust dimmer (see pages 74-75).
	Ballast defective.	Replace ballast as last resort and only if there is clear evidence of failure (smoke from fixture, acrid odor, fixture too hot).
Lamp flickers or blinks on and off.	Newness (normal for a short period with some new lamps).	None required.
	Low line voltage.	Check with local utility company if low line voltage is suspected.
	Temperature below 50 degrees Fahrenheit in lamp location, or cold draft on lamp.	If condition is permanent, shield lamp from draft or install low-temperature ballast.
	Starter defective.	Replace starter.
	Lamp-ballast mismatch.	Make sure lamp is type specified by fixture manufacturer.
	Lamp not properly seated in holders, or lamp pins bent.	Remove lamp. Check pins. If necessary, straighten pins with pliers. Clean pins with steel wool. Make sure lamp is properly seated in holders.
Short lamp life.	Specified lamp life is average. Some lamps will fail early.	None required (some lamps will burn longer than average).
	Lamp turned off and on frequently.	Avoid turning lamp off and on for short periods.
	If lamp fails within a few hours, ballast is wrong for fixture or is incorrectly wired.	Correct wiring or replace ballast.
	In some rapid-start and instant-start two-lamp fixtures, when one lamp burns out, the other lamp will dim or fail.	Replace burned-out lamps immediately. Substitute a good lamp for each lamp in the fixture, one at a time. When burned-out lamp is replaced, both lamps will light.
Tube ends discolored.	Normal near end of lamp life (see page 40).	Replace lamp.
	If lamp life is short, poor contact between pins and socket.	Clean and straighten lamp pins. Clean or replace lamp holder.
	Discoloration consisting of brownish rings at one or both ends may occur during normal operation.	None required.
Dark lengthwise streaks on lamp.	Condensed mercury on the lower part of the lamp; occurs when lower part is cooler than upper part.	Reposition lamp so that dark streaks are at top.
Excessive hum from the fixture.	Ballast causes hum. Some hum is normal. Hum will increase if ballast is loose or is overheating.	Turn off power. Open fixture. Check that ballast is securely mounted. Check that all connections are tight. Check that ballast is marked as correct for that fixture. Make sure ballast is properly ventilated. See page 40 for more on noise level.
Color variations in lamps of same type.	Some variation in color is normal.	None required.
	Significant age difference in lamps.	Substitute new lamps.
	Lamps operating at different temperatures.	Check for drafts. Equalize temperature as much as possible.
Radio interference (audible buzz).	Lamp radiation being picked up by radio antenna.	Separate radio and fixture as much as possible. Radio and antenna lead should be at least 8 feet from 40-watt fluorescent lighting and 20 feet from high-wattage tubes. Place metal screening (decorative screening may be used) around the fixture. Ground the screening to the fixture ground. Starter or ballast may be defective. Replace starter or ballast.
	Fixture radiation being transmitted through 120-volt power line.	Install a radio-interference filter between the fixture and the 120-volt power source. Both wired and plug-in type filters are available.

Doorbells and chimes

Doorbells and chimes operate on a much lower voltage than other home electrical devices. The operating voltages range from 10 to 20 volts; 16 volts is the most common. This low voltage is obtained by reducing the normal 120-volt power to the required 10- to 20-volt level. The device that provides the drop in voltage is a transformer. The transformer used for doorbell and chime circuits is a small apparatus—roughly the size of a softball. It has four electrical connections: two wires and two screw terminals. The wires are black and white and connect to a 120-volt power source. The screw terminals are the low-voltage output.

Normally, the transformer is located in the basement, close to a power source. The rest of the circuit—connected to the screw terminals on the transformer—is low voltage. Low-voltage wiring may be done with lightweight wire. No. 18 or 20 insulated wire, known as bell wire, is commonly used. There is no danger of serious shock or fire from low-voltage wiring.

In addition to the transformer, the doorbell or chime circuit consists of one or more push buttons at doors and, of course, the bell, buzzer, or chime unit. The low-voltage wiring may be routed through walls and behind baseboards and moldings with no special covering or protection. All the techniques shown elsewhere in this book for routing cable between floors and through walls may be used to route bell wire.

The diagrams opposite show the most frequently used installation circuits. Use these diagrams to make new installations or repair old circuits.

Installing the transformer. The connection of the transformer to the 120-volt power source is the only part of doorbell-circuit wiring covered by most electrical codes.

Turn off power at the service panel. Select a convenient wall or ceiling box for power tie-in near the planned transformer location. Connect the new transformer power cable to a 120-volt source cable in the wall or ceiling box. Terminate the new cable at a junction box. Remove a second knockout from the junction box and mount the box at the point where the transformer will be mounted. In basement areas the box and transformer are usually mounted on a joist.

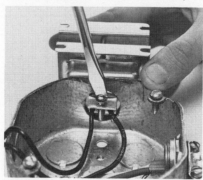

Put the threaded projection of the transformer (on the high-voltage side where the black and white wires are) through the knockout in the box. Secure the transformer to the box with a locknut. Connect the transformer wires to the power wires. Connect black to black and white to white. Be sure to connect the source-cable grounding wire to the junction box.

Connect the low-voltage wires to the screw terminals on the transformer. It will be easier to wire the rest of the circuit if the low-voltage bell wires are different colors. One color wire can be routed directly to the doorbell unit or units. The other color can be used to wire the push button or buttons to the doorbell unit.

Mounting the button and bell. Doorbell push buttons are mounted with two wood screws. You need only route the bell wire to the door, drill a ⅝-inch hole at the unobstructed button location, and fish the bell wire out through the opening. The wires are connected to screw terminals on the button. The button is secured to the door frame with wood screws. Whenever possible, situate the button under an overhang and out of the rain so the metal contacts on the button won't corrode. Route the bell wire through the wall to come out directly behind the bell, buzzer, or chime unit, following the manufacturer's instructions.

Troubleshooting doorbells and chimes

Bell does not sound.

Cause: Defective push button.
Correction: Remove push-button mounting screws. Loosen one screw terminal and remove wire. Touch wire to other terminal. If bell sounds, replace push button. If bell does not sound, check transformer and bell unit.

Cause: No power to transformer, or defective transformer.
Correction: Use voltage tester to check for 120-volt power at transformer wires. If OK, check for voltage at transformer low-voltage terminals. Use a low-voltage tester or voltmeter. If voltage does not register, replace transformer.

Cause: Defective bell unit.
Correction: If transformer is OK, connect two wires directly from transformer to bell unit. If bell does not sound, replace bell unit. If chime units do not sound, check for free movement of plunger. If plunger sticks, it can be cleaned with a small brush dipped in lighter fluid or silicone spray. Do not oil plunger.

Cause: Defective wiring.
Correction: If push button, transformer, and bell unit check OK, defective wiring is probable. Rewire the circuit. Use old wires to pull new ones through floor and wall openings.

Bell not loud enough.

Cause: Wrong combination of transformer and bell unit.
Correction: Check voltage required for bell unit. It is marked on unit near the screw terminals. Check voltage output marked on transformer. Replace either transformer or bell to get proper match.

Cause: Loose wire or corroded terminal.
Correction: Check terminals on transformer, push button, and bell unit. Clean, tighten, or replace as necessary.

Bell sounds continuously.

Cause: Shorted push button.
Correction: Check push-button contacts. Clean contacts or replace push button.

How doorbells are wired

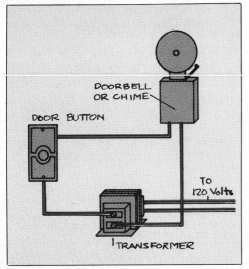

Two-terminal doorbell, chime, or buzzer wired for control from a single button.

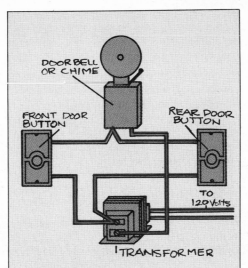

Two-terminal doorbell or chime wired to allow two buttons to ring one bell.

Two-terminal doorbell or chime and two-terminal buzzer wired for separate control.

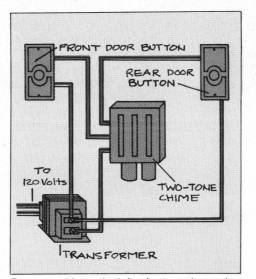

Two-tone chime wired for front- and rear-door control—two tones for front door, one tone for rear.

Two-tone chime wired for three-button control. If the chime has three terminals, the two-tone terminal can be wired to the front door and the single-tone terminal to the other doors.

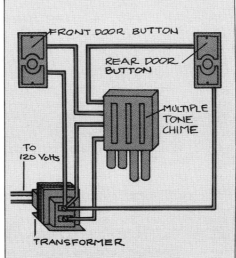

Multiple-tone chime wired for a two-button control—multiple tones for front door, single tone for rear door. Connect wires as marked on terminals.

What it takes

Approximate time: 1 to 2 hours.

Tools and materials: Bell, buzzer or chimes, one or more door buttons, bell transformer, junction box, cable clamps, cable, screwdriver, wire stripper or knife, solderless connectors.

Bats in the belfry

I bought a set of chimes that gave off a lovely, loud sound when I tested them at the hardware store. But when I wired them into the transformer on my old doorbell system, they gave off a dull, feeble "bonk" you could hardly hear. After I made a couple of trips to the store and back, the dealer asked what kind of transfomer I had. It turned out that the old transformer could handle the doorbell but was the wrong match for the new chimes. Copy the specs from the old transformer and bring them with you when you go to buy new chimes, or be prepared to make a second trip for a new matching transformer.

Practical Pete

How to check the sound level of a new doorbell location.

After the transformer has been installed, temporarily connect two long lengths of bell wire to the transformer low-voltage terminals. Connect one of the wires to one terminal on the bell, buzzer, or chime. Turn on power to the transformer circuit. Then hold the doorbell at or close to the planned location. Briefly touch the other wire from the transformer to the other terminal of the doorbell. This will cause the bell or chime to ring. Have a helper go around the house to check that the bell is audible. A change in location of only a few feet can often make a big difference in the sound level.

Troubleshooting branch circuits

If trouble develops in part of your home electrical system, you can usually determine the cause and, in most cases, correct the trouble by making a series of inspections and tests.

Two types of branch-circuit failure can occur: no power in one or more circuits or a short in one circuit. The short will cause a circuit breaker to trip to OFF as soon as you reset it, or it will cause a fuse to blow as soon as main power is turned on. Either condition may occur due to failure in your house power cables. Failure within cables, however, is extremely rare. Breaks in wires and even shorts between conductors may occur in *new* wiring. But they seldom appear in wiring that has been satisfactory for a period of time *unless a cable is damaged*. When power failures or shorts occur, make an especially careful check of all wiring and circuit boxes near recent repairs or renovation.

No power—circuit breaker not tripped, fuse not blown

Complete power failure in your home is probably due to failure in utility power lines. Report this type of problem to your utility company.

There are three types of *partial* power failure:

Loss of power on several circuits—like full power failure—is most likely to be a problem that must be corrected by your utility company. If your house has three-wire service (most homes do), failure can occur in half of the utility-supplied power. Report this type of failure to your utility company.

Loss of power on only one circuit appears to be the same as a tripped circuit breaker or blown fuse. If a check reveals that the circuit breaker is not tripped, or the fuse is not blown, the power loss must have occurred between the service panel and the first outlet, junction box, or ceiling box that's on the circuit. Set the circuit breaker to OFF, or remove the fuse, and make the circuit inspection described below. If there are no loose or broken wires, the trouble is probably in the service panel. The circuit breaker may be defective or the fuse contacts broken. Working inside a service panel can be dangerous. Your best bet is to let a licensed electrician find and correct the trouble.

Partial loss of power on one circuit can almost always be traced to a defective switch or outlet or an open connection in a box. With circuit power turned on, use your voltage tester to determine exactly which parts of the circuit have power and which do not. Next, turn off power to the circuit. Make a careful inspection of the wiring in each electrical box that appears to have no power. Look for loose connections on switches and outlets. Check that solderless connectors are in place and are tight. Check for broken wires at or near connections. Conductors can be nicked when insulation is removed. Additional strain is put on the conductor when it is bent around a screw terminal. Breaks can occur at these points.

Short circuit—breaker trips or fuse blows immediately

Follow the procedure described on page 18. If all plug-in appliances and lamps on the failed circuit have been checked and found to be OK, the short circuit must then be in the circuit wiring.

Visual check. With the circuit breaker off, or the fuse removed, take off faceplates and check wiring in all switch and outlet boxes. Keep your voltage tester handy and remember to test before you touch any bare wires. *When a circuit is defective, power may be present where it normally would not be.*

In each box, look for broken or exposed conductors or missing solderless connectors. If wiring appears OK, disconnect and check switches. Use your continuity tester and follow the procedure on page 36. If the tester lights when it should not, the switch has an internal short and should be replaced. Shorts in outlets are rare but can happen. Remove the outlet. Connect the alligator clip on the continuity tester to a brass terminal. Insert the probe in all slots. Touch the probe to the mounting bracket and, if the outlet has one, to the green-tinted grounding screw. The tester should light only in one top slot and one bottom slot. If the connecting metal tab between the brass terminals has been removed, repeat the test with the alligator clip connected to the other brass terminal. In this case the tester should light in only *one* slot during the test of each brass terminal.

Ceiling boxes and junction boxes are harder to get at than wall outlets, so leave them until last. Make the same inspection described above for wall boxes. The canopy on a ceiling fixture must be loosened and the fixture lowered for you to check the wiring inside a ceiling box (see pages 38 and 39). Do not allow fixtures to hang by their electrical connections.

Electrical check. If the cause of the short circuit cannot be determined by the visual check, electrical tests should be made to locate the trouble. The general procedure for the electrical check consists of disconnecting portions of the circuit and then turning on power to the remaining portions. At the point where the circuit breaker no longer trips—or the fuse does not blow—when power is again turned on, the disconnected portion of the circuit contains the short.

Once you have isolated the trouble to a small part of the circuit, the job of locating the defect becomes easier. Detailed continuity checks can be made for every device on the defective portion of the circuit. Check cabling with your continuity tester. Connect the alligator clip to a black wire. Touch the probe to all white wires and to ground wires in the box. The tester should not light. (Remove all light bulbs from fixtures on the defective portion of the circuit when checking cabling.) The procedure for making electrical checks at the main test points in a branch circuit is described opposite. Adapt it as necessary to fit test points available in your particular circuit.

Junction box

The junction box serves as a tie point for two sections of the circuit. Be sure power is off at the service panel and test all wires before you touch them. Remove solderless connectors and separate all wires. No wires should touch each other or any metal.

Turn on power at the service panel. If the circuit breaker trips to OFF or the fuse blows, the short must be in the service panel or in the wiring between the service panel and the junction box.

If the circuit breaker stays on or the fuse does not blow, the wiring from the service panel to the junction box is OK. At the junction box, use your voltage tester to determine which black and white wires are connected to the service panel. Touch the tester probes to each pair of wires until the tester lights. Do not touch bare wires.

Turn off power at the service panel. Connect the black and white wires from the service panel to one of the other pairs of wires, using solderless connectors. Turn the power back on at the service panel. If power stays on, the portion of the circuit connected to the service-panel line is OK. In the diagram at right, for example, if pair A is connected to pair B, and pair C is left unconnected, the short would have to be in the portion of the circuit powered by cable C.

Middle-of-run ceiling box

Turn off power to the circuit. At a middle-of-run ceiling box, disconnect and separate all cables. Disconnect the fixture. At the junction box, connect source-cable A to source-cable B. Turn on power. If power stays on, either the fixture or one of the cables connected to source-cable B contains the short.

Turn off power. Connect one cable to the source cable. Turn on power. If, for example, cable D is connected to source-cable B and power stays on, the short is either in the fixture or in the loop switch.

Turn off power. Connect the black-coded white wire from the loop switch to the power cable. Turn power on again. If power stays on, the short is in the ceiling fixture. Check it with your continuity tester as described under "Testing the parts of a lamp" (page 34).

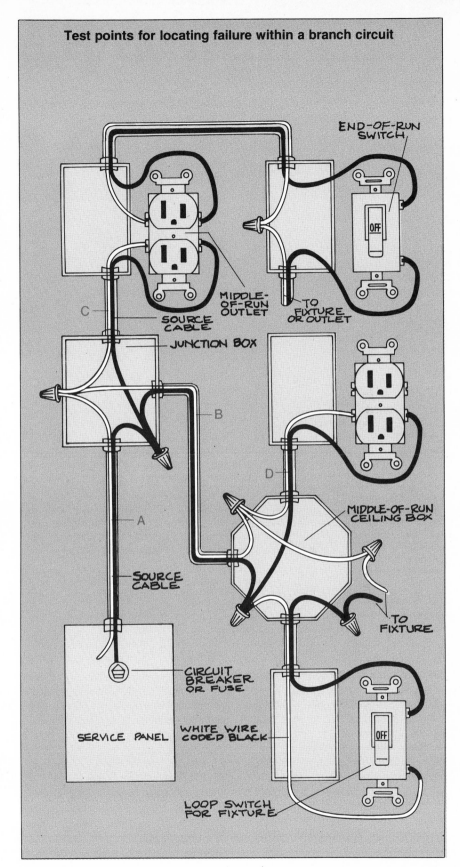

Test points for locating failure within a branch circuit

END-OF-RUN SWITCH

MIDDLE-OF-RUN OUTLET

TO FIXTURE OR OUTLET

SOURCE CABLE

C

JUNCTION BOX

B

D

MIDDLE-OF-RUN CEILING BOX

TO FIXTURE

A

SOURCE CABLE

CIRCUIT BREAKER OR FUSE

SERVICE PANEL

WHITE WIRE CODED BLACK

LOOP SWITCH FOR FIXTURE

Middle-of-run wall outlet

A middle-of-run wall outlet is a good point at which to disconnect a portion of the circuit. With power off, disconnect and separate all wires. Turn on power. If power stays on, the short is in the portion of the circuit past the end-of-run outlet, as shown in the diagram above.

4.NEW WIRING

As your family grows and matures and as your life-style changes, your home wiring system should change too. When receptacles become crowded with plugs and a tangle of extension cords is concealed behind a sofa or chair—or when you want to make an attic livable—you can make your system safer and more convenient yourself by adding new wiring, outlets, and fixtures to the circuits in your existing wiring.

Advance planning. When you install a cir-cuit extension, two important points must be considered: First, the additional electrical load that will be carried by the new wiring must be added to your present circuits in a way that will avoid overloads. Second, the outlets, switches, and cables you will install must be planned to fit the construction of your house. Advance planning in this area will make it possible to do a neat, professional job with minimum effort and a maximum of satisfaction.

Which circuit to add onto?

Check the circuits that are available in the area of your house in which the circuit extension is needed. Usually you will find that more than one circuit can be used. Your main service panel shows the voltage and amperage rating of the circuit breaker or fuse that protects each available circuit. Multiply the amperage rating of the circuit breaker or fuse by the voltage of the line (120 or 240) to determine the maximum wattage that the circuit can carry.

For example, a 15-amp circuit breaker or fuse on a 120-volt line can carry 1800 watts; a 20-amp circuit breaker or fuse on a 120-volt line can carry 2400 watts.

Next, calculate the present load on the circuits. To do this, add up the wattage of all lights on the circuit and all appliances that are plugged into receptacles on the circuit. The difference between the present circuit load and the maximum circuit wattage is, in principle, the amount that can be added. But it is wise to leave a margin of safety in this calculation in case you later want to add a new appliance or replace an existing one with another that consumes more watts. On a 120-volt line, for example, try to limit the 15-amp circuit to 1500 watts and the 20-amp circuit to 2000 watts. Remember, too, that appliances containing motors of one-quarter horse-power or more (humidifiers, fans, small air conditioners) always draw heavy current when starting. The margins of safety noted above will allow for this. The motors in phono turntables and tape players are quite small and need not be given special consideration. Use the wattage rating on the manufacturer's nameplate.

If you wish to add a larger appliance (large air conditioners, room heaters, stoves, dryers), a new 240-volt circuit, or even an increase in the service provided by your utility company, may be required. In either case, these changes should be made by a licensed electrician.

If a new circuit is needed, you can keep the cost to a minimum by having an electrician wire the new circuit from your circuit-breaker or fuse panel to a convenient junction box. You can do the wiring from there to the outlets you need.

Evaluating available power

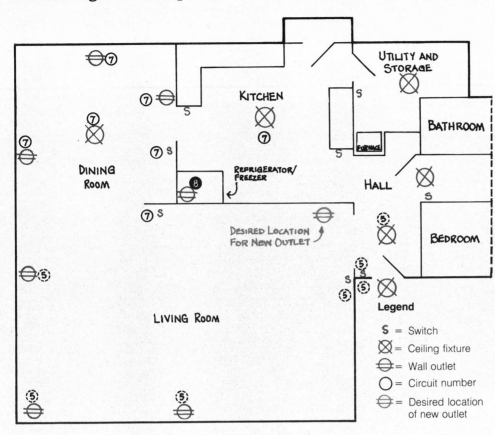

KITCHEN

UTILITY AND STORAGE

BATHROOM

DINING ROOM

REFRIGERATOR/ FREEZER

HALL

DESIRED LOCATION FOR NEW OUTLET

BEDROOM

LIVING ROOM

Legend

S = Switch

⊗ = Ceiling fixture

⊖ = Wall outlet

◯ = Circuit number

⊜ = Desired location of new outlet

The best source of information for planning circuit extensions is a circuit map. The procedure for preparing a circuit map of your home is explained on page 19. The map shown here is a typical example. It is referred to in the text and in the circuit-analysis chart below.

In this example, let's assume the new outlet you are planning is to be used for a console-type color-television set. The manufacturer's nameplate on the TV indicates that the power rating of the set is 280 watts.

All three circuits appear to have sufficient capacity available for this extra load. It would be unwise, however, to connect the new outlet to circuit number eight. At first glance it would appear that there is sufficient capacity (1250 watts for a 280-watt load) and a convenient location on the opposite side of an interior wall.

The difficulty in this case would arise from the heavy surge of current that occurs each time a large motor, such as that used in the refrigerator/ freezer, is started. This surge would cause an annoying shrinkage of the TV picture and loss of synchronization for several seconds. A momentary circuit overload would also occur. The choice then is between circuits five and seven. Since circuit five has greater capacity available, it should be the first choice, but either circuit could handle the additional load. The next step in planning the circuit extension is to check the electrical boxes on circuits five and seven to find the easiest place to connect the new power cable. The decision on where to make the connection should also take into account the possible ways of routing the new cable. The next few pages give you the information you need to complete the plan.

How to calculate unused capacity in circuits potentially available for extensions			
Circuit available	**No. ⑤ circuit (15 amps)**	**No. ⑦ circuit (15 amps)**	**No. ❽ circuit (20 amps)**
Total power available	1800 watts	1800 watts	2400 watts
Less safety margin	300 watts	300 watts	400 watts
Less present load*	650 watts	1050 watts	750 watts
Capacity available	**850 watts**	**450 watts**	**1250 watts**
*Analysis of present load	Outside light ———— 100 watts Entrance light ———— 100 watts Living-room lamps (3) ———— 450 watts Total ———— 650 watts	Dining-room ceiling fixture ——600 watts Kitchen ceiling fixture ——150 watts Dining-room stereo outlets ——300 watts Total ———— 1050 watts	Refrigerator/ freezer ———— 750 watts

Where to join into a circuit

To decide where to connect into an existing circuit, you must be able to identify the job each wire is doing inside each box on the circuit. Various kinds of wall and ceiling boxes are used on general-purpose circuits. All of the types that you are likely to encounter are shown on page 24. To determine the function of a box and where it is connected electrically in the present circuit wiring, compare the connections in the box with the descriptions on these pages. On page 24, you will find the recommended maximum number of wires for each box. (See page 22 for additional information.)

To open ceiling boxes

1. Turn the fixture on by setting the wall switch that controls the light to the ON position.
2. Turn off power at the service panel. The fixture should now be off. Leave the wall switch in the ON position.
3. Remove the screws or nuts that hold the fixture to the ceiling. Pull the fixture out of the box to expose the wires to view.

4. Support the weight of the fixture with one hand while you unscrew the wire nuts (page 27). The weight of the fixture should now separate the wires; if not, grasping them by the insulation, pull each set apart.
5. Set the fixture aside and loosely reassemble the hardware so you won't forget how it goes together.

The receptacle diagrams show three-prong-type outlets with grounding wires attached. On page 51, for the sake of simplicity, grounding wires are not shown. Grounding wires must be connected when the new cable is installed. Illustrations on pages 52 and 53 show how to connect grounding wires.

To open wall boxes

1. Before you start taking any electrical box or fixture apart, turn off the circuit breaker or remove the fuse that protects the circuit to be checked.
2. Remove the cover plates from the wall boxes that are most convenient for connecting the circuit extension you wish to add. Remember to *test before you touch* any exposed wires or terminals.
3. To see how the box is wired, you may have to loosen the screws that secure switches and receptacles to the boxes, and pull the switch or receptacle all the way out.

How to distinguish typical wiring functions

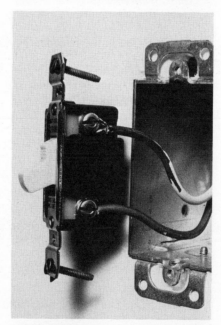

Loop switch. The wall switch shown here is called a loop switch because it completes the hot-wire circuit to a fixture. It can be identified by the black coding added to the white wire (paint or tape put on the ends of the white neutral wire when it was used as a hot wire). Loop-switch receptacles contain no true neutral wire, so you cannot make a connection to the circuit at that point.

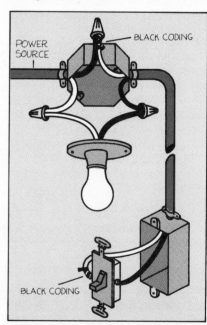

Case number 1: End-of-run ceiling fixture with loop-switch control. This ceiling fixture is the last device on the circuit (end-of-run) and is controlled by a wall loop switch. You can connect into the line only at the ceiling box. This is easy if the box is accessible from an attic or crawl space. If a larger box is required, you can install it from the attic or crawl-space side without damaging the finished ceiling.

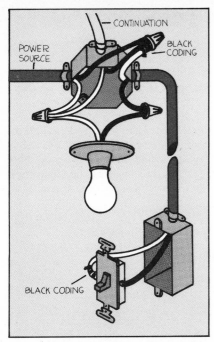

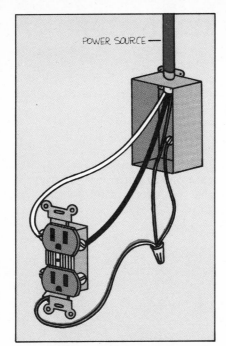

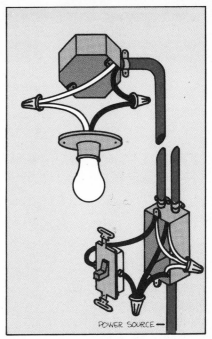

Case number 2: Middle-of-run ceiling fixture with loop-switch control. This is similar to case number 1, except that circuit power continues beyond the fixture. Because of the extra cable to the rest of the run, the middle-of-run ceiling box is likely to become overcrowded if an additional cable is added. As in the case of the end-of-run ceiling box, the ease or difficulty of box replacement depends on box location.

Case number 3: End-of-run receptacle. If the receptacle is *not* controlled by a switch this is the ideal point at which to tie in new wiring. The box is uncrowded. Only one cable is in the existing box. The new cable can easily be brought into the box and connected to the spare receptacle terminals.

Case number 4: Middle-of-run switch. As in case number 2, overcrowding of the box is likely, and enlarging the box will require considerable wall repair. Unless you are planning to install paneling or paper the walls, this is not a good choice for tie-in.

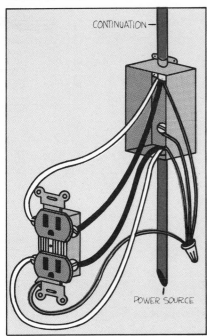

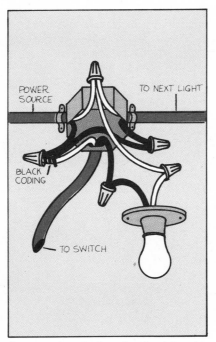

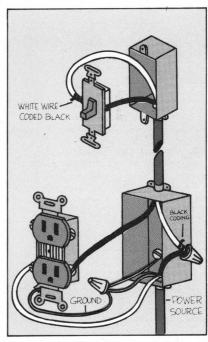

Case number 5: Middle-of-run receptacle. If the box is large enough and the receptacle has no switch, tie in at this point. You should not connect more than one wire to screw-type terminals, and jumpers might overcrowd the box. Substitution of a larger box or ganging the existing box with another can be a problem, as noted in case number 4. If so, look for another tie-in point.

Case number 6: Junction box. Junction boxes are good tie-in points if they are reasonably accessible. CAUTION: Junction boxes may be used for more than one circuit. Don't assume all power is off because one circuit breaker is off or one fuse has been removed. TEST BEFORE YOU TOUCH.

Case number 7: Switched end-of-run receptacle. Electrically, a switched receptacle is similar to a switched ceiling fixture. Since a switched receptacle is wall mounted, cable routing is usually easier. But you must take special care when making connection to a switched receptacle. See page 53 for the difference in connection, depending on whether the added circuit is, or is not, to be controlled by the switch. Check box size. Overcrowding of the box may be a problem.

Planning where to run new cable

At this point you have checked the electrical boxes available for the circuit extension, and you have decided which box is the best place to make the connection to your existing system. If possible, at this stage in the plan, have more than one tie-in box in mind. The best way to route the new cable depends mostly on the purpose of the extension. Cable-routing ideas for specific installations are included in the detailed instructions in the balance of this chapter. There are, however, certain points that you should keep in mind when planning the best way to route cables for any installation. These points are related to common structural factors found in most houses built within the last thirty years. The illustration below shows a typical frame construction.

Places to run electric cable

A. Interior walls and ceilings are easiest to work with. The space between wall surfaces is usually empty—except for existing wiring—and clear of obstructions.
B. In multifloor homes with basements, it is usually best to make first-floor runs along floor joists, accessible from the basement.
C. .Second floor runs are best made along second-floor ceiling joists, accessible from the attic or crawl space.
D. Exterior walls are difficult to work with. They may be solid masonry (or concrete block), with no hollow spaces for routing wires and sinking boxes. Even if they are made of wood and framed like interior partitions, the spaces between studs are likely to be insulated, are almost certain to be interrupted by fire-stops (horizontal wood pieces inserted between studs to block the spread of flame). They may also have diagonal braces running across them. If you must extend wiring across an exterior wall, consider surface wiring. See page 84.
E. To make a run between basement and attic or crawl space, check the plumbing-vent-pipe area. This pipe runs from the basement sewer line up through the house to the roof. The space provided for the pipe is usually clear, fairly large, and unobstructed from the basement to the attic. It is much easier to drop a cable through this area than to penetrate walls and floors.

Instructions for carpentry required to run cable in various locations begin on page 62.

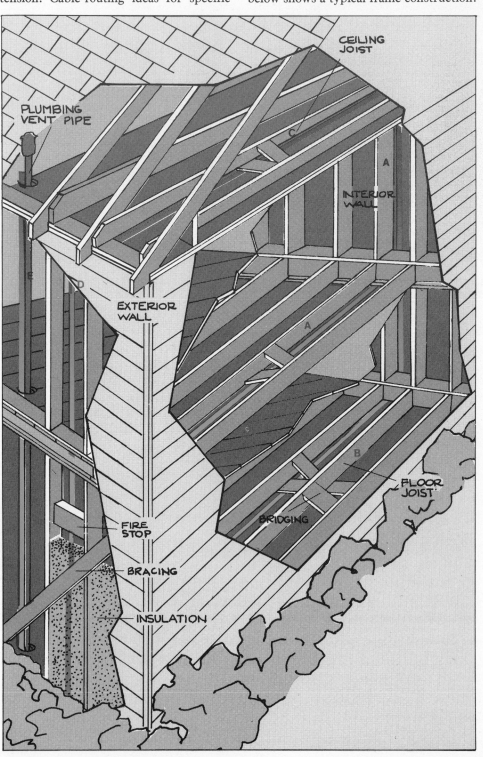

How to identify wires in electrical boxes

When you have decided which box you will use to connect your new cable to the existing wiring, you have to identify and tag the wires in the box. You need to know which wires go to the power source. Your new cable will be connected to the source cable. To avoid wiring errors, it is necessary to know which wires go to the rest of the run (if the box is middle-of-the-run) and which are switch wires. The continuity tester is helpful in checking switch wiring.

Check the testers before using them. The voltage-tester bulb should light when the probes are inserted in the slots of a "live" receptacle. The continuity-tester bulb should light when the probe and alligator clip are touching.

When you have identified the source cable and, if present, the switch cable, you know all you need to know about the wires in the box. Cables in the box other than source cable and switch cable are cables to receptacles, outlets, switches, or ceiling or junction boxes on the same circuit.

The red-faced TV

I was *sure* there was something wrong with the brand-new TV set we bought for our redecorated living room. It flickered and acted sick at the oddest times. I chewed out the dealer, who picked it up and returned it an hour later. "Perfect," he said. "Oh, yeah?" sez I, and plugged it in. It worked fine just then, but it acted up again that evening. After another trip, the serviceman asked where my main fuse panel was. He located the circuit the TV was on and discovered that our dishwasher was also on that circuit. "It's the motor in the dishwasher, Pete," he explained wearily. "When it starts up, it draws extra current which makes the TV flicker. Take your choice, buddy; plug the TV into another circuit without a motor-driven appliance on it, or wash your dishes by hand."

Practical Pete

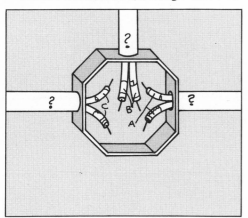

Source cable

1. Turn off all power to the box.
2. Separate all conductors that were joined by wire nuts. To simplify reconnecting the wires, tag wires to show which were joined. Use colored plastic tape or tape that can be marked with ball-point pen. Code all wires joined by one wire nut with one letter. Code the first group "A," the next group "B," and so forth.
3. After tagging, spread out all the wires in the box so that the bare conductors do not touch each other, the box itself, or any nearby metal surface. It is important to do this carefully to avoid a short circuit between black and white conductors or between a black conductor and ground.

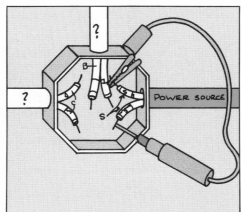

4. Use alligator clips to connect your voltage tester to a black and a white wire entering the box from a single cable, as shown above.
5. Turn on power at the service panel. If the tester bulb lights, the cable the tester is connected to is the source cable. If the bulb does not light, turn off power at the service panel. Remove the tester from the first pair of wires and connect it to the wires of another cable entering the box. Repeat the procedure. Continue until you locate the source cable.
6. When you have identified the source cable, turn off power and mark the tags on the source-cable wires with an "S" or other code letter. You will connect your circuit extension to these two wires, either directly or by means of jumpers, as shown.

Testing for power in wired boxes

Junction boxes may contain cables on more than one circuit. Always test before you touch.
1. Turn off power at the service panel.
2. *Carefully* remove wire nuts from all black wires or black-coded wire connections in the box. Use one hand and touch only the plastic portion of the wire nut.
3. Touch one probe of the voltage tester to the exposed black-wire conductors. Touch the other probe to the box. If power is off, the bulb will not light. Repeat this test for all black-wire and black-coded wire connections in the box.
4. Carefully remove wire nuts from all white-wire connections.
5. Touch one probe of the voltage tester to a white-wire connection. Touch the other probe to all black-wire and black-coded wire connections in the box. If power is off the bulb will not light. Repeat this test for all white-wire connections.

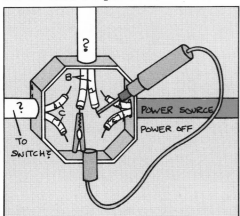

Switch wires

If the box you are working with has a switch connection, the loop cable to the switch can be easily identified with the continuity tester.
1. Turn off all power to the box.
2. The wires in the box should be disconnected and spread, as in step 2 above.
3. Set the wall switch to ON.

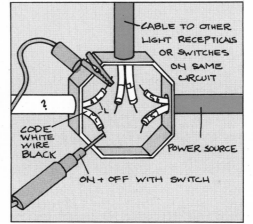

4. Touch the continuity-tester probe and alligator clip in turn to each pair of black and white wires.
5. When the tester lights, have an assistant flip the wall switch on and off a few times.
6. If the tester light goes on and off as the switch is moved, mark those wires "L" for loop.
7. If the tester light remains on, regardless of the switch position, continue checking other pairs of wires until you find the pair controlled by the switch.

Connecting new cable

Remember, if you are working with a junction box containing cables on more than one circuit, you will have to turn off two circuit breakers—or remove two fuses—to turn off all power in the box. In this case there will be two source cables. You can identify both by following the procedure given on page 51. When the voltage tester is connected to the wires in the box, turn on power to one circuit at a time. In this way you will know which source cable is connected to which circuit breaker or fuse.

Also, grounding wires of all cables entering a box should be joined. If plastic-sheathed cable is used, a jumper should be connected from the (usually uninsulated) grounding-wire connection to the metal box. If steel-armored cable is used, the box is grounded automatically when the cable ground wire, the cable clamp, and the clamp mounting are installed.

What it takes

Approximate time: 1 to 2 hours.

Tools and materials: Plastic-sheathed, or armor-covered cable, cable clamps, screwdriver, wire stripper or knife, solderless connectors.

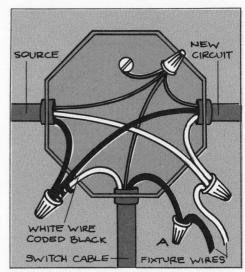

Connecting new cable to end-of-run ceiling fixture with loop-switch control. When the new circuit is added, two of the connections will join three wires, rather than two. Be sure to use large-size solderless connections (wire nuts). As shown, the new circuit will not be controlled by the loop switch. If control by the switch is desired, connect the new-circuit black wire with the switch cable and fixture wire at point A.

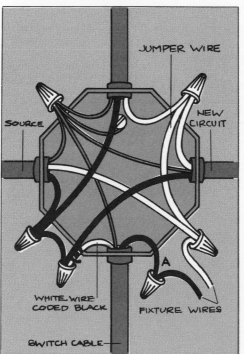

Connecting new cable to middle-of-run ceiling fixture with loop-switch control. Note that jumper wires are required to assure a good conductor connection with wire nuts. Bare grounding wires may be joined as shown. As with the end-of-run ceiling fixture with loop-switch control, if you want to control the new circuit from the present ceiling light switch, connect the new-circuit black wire at point A. Assuming that the box shown also contains a mounting stud for the ceiling fixture, the box would have to be at least 2⅛ inches deep to accommodate the additional connections.

Connecting new cable to end-of-run receptacle. This is the easiest connection to make. The outlet shown is a three-prong grounded type, not under switch control. The new circuit is connected to the spare screw terminals at the outlet. Be certain both black wires are connected to the brass-colored terminals on the receptacle.

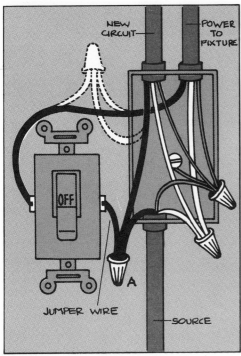

Connecting new cable to middle-of-run switch. To provide switch control of the new circuit in this instance, connect the new-circuit black wire, the black wire to the fixture, and a jumper to the switch, as shown by the dotted lines. In this case the connections at point A would not have to be made. The black wire from the source would remain directly connected to the switch.

Connecting new cable to middle-of-run receptacle. Screw-type terminals should never be used to join wires. Only one wire should be connected to each terminal. Make additional connections with jumper wires, as shown, or by using spare terminals.

Population explosion
Have you ever tried to carry two mean dogs, a strange alley cat, and a bantam rooster in the back of your station wagon?

I picked the perfect wall box to run my new wiring out of. Opened it up and it was already pretty full. Well, trying to connect the new wiring onto that mess of connectors and jam it all back in the box was only slightly easier and quicker than performing dentistry on a diamondback rattler in a gunnysack full of his relatives.

Practical Pete

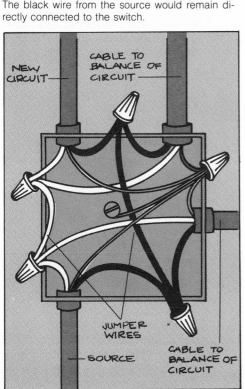

Connecting new cable to junction box. This junction box contains wiring for only one circuit. Remember, junction boxes may contain wiring for more than one circuit. Test before you touch.

Connecting new cable to receptacle with loop-switch control. If the new-circuit black wire were connected to the spare brass-colored terminal on the outlet, the new circuit would be controlled by the same switch that controls the outlet.

5.SWITCHES

What switches do electrically

The simplest and most common switch used in home wiring is the Single-Pole, Single-Throw (SPST) type. This switch completes or interrupts the hot-wire circuit. Electrically, it acts just like the old-fashioned knife switch (page 55, far right). SPST switches have two terminals.

Another type of switch is used in circuits that allow lights or outlets to be controlled from two locations. This type is the Single-Pole,

Double-Throw (SPDT) switch. This switch allows a conductor connected to the center (or common) terminal to be connected to either of two wires. SPDT switches have three terminals and are sometimes known as three-way switches. Don't let this name confuse you. Remember, these switches can be used only to control a light or outlet from *two* locations, and the switches have only *two* positions.

When you want to control lights or

outlets from three or more locations, a third type of switch must be used. This is referred to as a special type of Double-Pole, Double-Throw (DPDT) switch. This switch has four terminals. Current can flow through the switch in both switch positions. The difference is that one position provides straight-through connections and the other position provides crossover connections. Connected pairs of terminals need no OFF.

Standard toggle switches

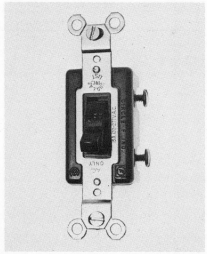

Single-pole, single-throw toggle switch. This is the type of switch most often encountered in house wiring. It has two brass-colored terminals. The switch positions are marked OFF and ON. Terminals may be located on the side, top or front of the switch.

Single-pole, double-throw toggle switch. Two things are special about this switch; it has three terminals, and it does *not* have OFF and ON markings. The three terminals usually consist of two brass-colored terminals and one copper-colored terminal. In all cases, one terminal (the common terminal) is darker than the other two.

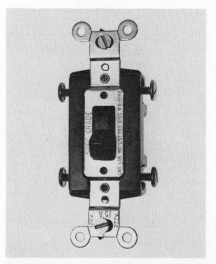

Special double-pole, double-throw toggle switch. This switch can be distinguished from the SPDT switch only by the number and color of terminals. The DPDT switch has four terminals all the same color. It, too, has no OFF or ON markings. There is always an electrical connection between pairs of terminals, and so the switch is never "off."

Special-function switches

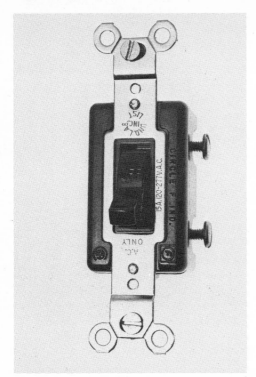

Quiet switch. This switch is mechanically designed to move from one position to the other with almost no noise. It is available in all electrical types. A quiet switch is slightly more expensive than a snap-action switch. In any installation where switch noise is objectionable, a quiet switch should be considered.

Locking switch. This switch must be unlocked with a key before power can be turned on. It is especially useful when you want to prevent unauthorized use of power tools or outside outlets. Locking switches with the UL symbol are designed so that metal objects inserted into the keyhole cannot come in contact with electrical parts.

Old-fashioned knife switches

Three common types are shown, below, to demonstrate how switches act to complete the electrical circuit. Because of their exposed terminals, knife switches should *never* be used in home power wiring.

SPST
Single-Pole, Single-Throw

SPDT
Single-Pole, Double-Throw

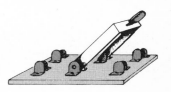

DPDT
Double-Pole, Double-Throw

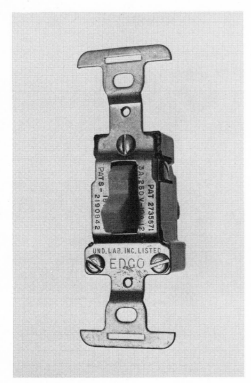

Time-delay switch. This switch contains a short-term timer (about 45 seconds). When the switch is turned off, the timer is started, and the light will remain on for about 45 seconds. This gives you enough time to get into the house or to another lighted area before the switch turns the light off.

Time-clock switch. This switch provides the same control as a plug-in timer, but it can be mounted in wall switch boxes. It can be used to control lights, air conditioners or other devices over a 24-hour period.

A toggle switch with four terminals *and* OFF and ON markings is sometimes used to control each 120-volt half of the 240-volt power to large appliances. This type of DPDT switch *cannot* be used for multiple-control installations.

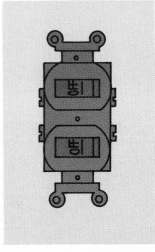

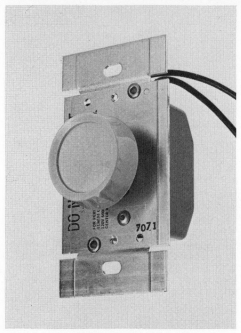

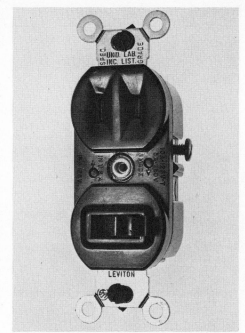

Dimmer switch. A dimmer switch combines switch control and brightness control in one package. A dimmer switch is slightly larger than a standard switch but can be installed in the same switch box. In addition to the usual switch markings, a dimmer switch has maximum-control wattage ratings. A common rating is 600 watts. This means that the total wattage of bulbs in the fixture controlled by the dimmer must not be greater than 600 watts.

Switch-receptacle combination. This device combines a standard SPST switch and one wall outlet in a single package. It provides an easy way to add a receptacle at any middle-of-run or end-of-run switch location. Loop switches contain no neutral wire conductor, and therefore a receptacle cannot be installed at that location. (For switch-receptacle wiring, see page 57.)

Who needs static?
Don't take it out on your TV set when interference is caused by a lighting-fixture dimmer switch on the same circuit. Dimmer switches are for lighting control only and can disrupt reception on TVs and radios. It's better to plug these appliances into another circuit. If you must use the same circuit, various kinds of interference suppressors, which may be added to the TV or radio power cord, are available. See pages 58 and 59 for dimmer-switch installation guidelines.

Practical Pete

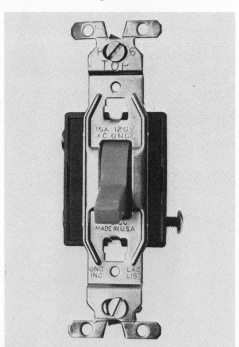

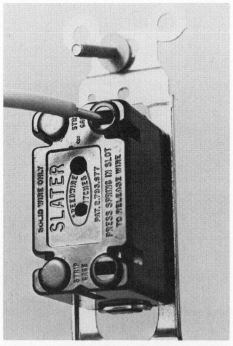

Silent (mercury) switch. This type of switch makes electrical contact by tilting a sealed container of mercury. When the mercury flows to one end of the container, it completes the circuit between two switch contacts. When tilted the other way, the mercury flows away from the contacts, and the switch is off. This action is completely silent and almost wear-free. A mercury switch is more expensive than a snap-action or quiet type but will last indefinitely. Because gravity controls the flow of mercury, this switch is marked TOP at one end and must be mounted vertically with that end up.

Push-in connection. Both switches and receptacles are available with push-in type connections. The back of the body of these switches or receptacles has a molded indentation called a strip gauge. This shows how much bare wire is needed to make good electrical contact. After you strip insulation from the wire, cut the bare conductor to the length of the strip. To connect the wire, simply push it into the terminal hole. To disconnect the wire, insert a small screwdriver into the adjacent release slot and pull the wire out of the terminal hole. These devices should only be used with copper wire.

Wiring switched receptacles

To gain an extra outlet quickly and easily, replace a standard toggle switch with a switch-receptacle combination. The switch to be replaced must be a middle-of-run or end-of-run switch. Refer to pages 48 and 49 for a description of how to identify the switch's position on the circuit. The switch-receptacle combination can be wired so that the new switch operates the same as the old one and the receptacle is always "live," or the new switch can operate the same as the old one and also control the receptacle, as shown in the two diagrams below.

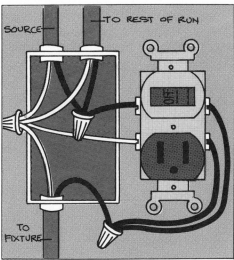

Switch-receptacle combination wired so receptacle is always "live."

Switch-receptacle combination wired so receptacle is switch controlled.

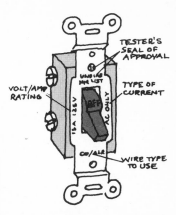

What the markings mean: Both switches and receptacles are marked to indicate how the device should be used. These markings are stamped into the metal mounting strap or are molded into the plastic body. The markings tell you the maximum voltage and amperage at which the device can be used. A typical marking is "15A-120V." This means that the total power that can be controlled by the switch or plugged into the receptacle is 15 amperes at 120 volts, or 1800 watts.

Pilot-light switches

Two types of pilot-light switches are available. One type is designed to glow in the dark so it will be easy to find at night or in dark locations. The other type has an indicator light to call attention to the fact that the switch is on. This is particularly useful when the light or appliance controlled by the switch is in a remote location, such as with outdoor lights or garage lights. Its energy-saving potential is significant.

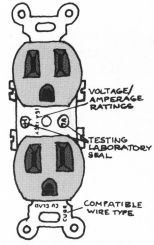

Acceptable wire types are indicated by abbreviations, as follows: "CU" or "CU-CLAD only" means either copper or copper-clad wire may be used (solid, uncoated aluminum may *not* be used). "CO-ALR," "CU-AL," or "AL-CU" means that solid aluminum, copper-clad aluminum or solid copper may be used. Listing by Underwriters' Laboratories, Inc. is shown by the UL symbol.

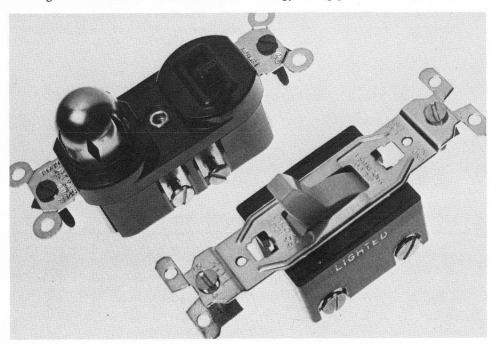

Incandescent dimmer switches

Dimmer switches provide continuously variable levels of light to suit the mood or use of an area or a room. When adjusted for less than full brightness, they also have the practical advantages of reducing power consumption and therefore cost, and of prolonging the life of the bulbs.

Any standard toggle switch that controls incandescent lights can be replaced with a dimmer switch. Dimmer control of fluorescent lights is a bit more complicated. (See page 74.)

The most common incandescent-light dimmer switch has a push-pull, on-off switch and a rotating brightness control.

Dimmer switches should be used only to control brightness of illumination. Any attempt to control appliances by means of a dimmer switch will result in severe damage to the appliance.

Dimmer switches are available in various wattage ratings. The rating applies to the total wattage of all lights controlled by the dimmer. A common rating is 600 watts, which would be more than adequate, for example, for control of a dining-room fixture containing six 75-watt bulbs.

Single-pole, double-throw dimmer switches are available for use on lights under two-switch control. Only one switch, however, can contain a dimmer. The level of illumination is controlled by the dimmer, but the light may be turned off or on from either switch.

Dim, dimmer, dimmest
My wife wanted me to replace the old toggle switch in the dining room with a dimmer switch. But I didn't realize that the big housing on the back of the dimmer would fill up almost all of the space in the wall box. In addition, I had to allow space for at least two wire nuts. Now, since my wall box was nailed to the studs, the nail heads stuck out so much that the housing wouldn't fit all the way inside the box and the faceplate wouldn't screw flat against the wall.

Measure the available space inside the wall box *before* you buy the switch.

Practical Pete

Dimmer switches are connected by wires rather than screw-type terminals. Standard (single-pole, single-throw) dimmer switches have two black leads. These are connected to the two black (or black-coded) switch leads in the box by means of solderless connectors (wire nuts).

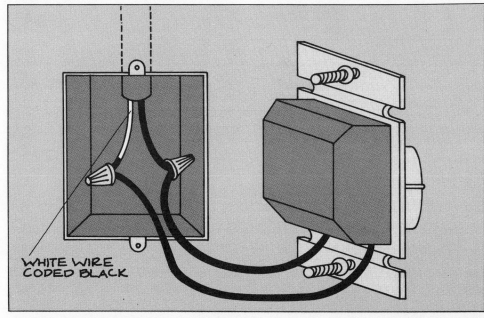

WHITE WIRE CODED BLACK

Single-pole, double-throw dimmer switches have three leads. Markings on the switch and the package indicate which is the common terminal. Wire the dimmer the same as the single-pole, double-throw switch that it replaced. If the dimmer is used in a new two-switch installation, wire the dimmer the same as the switches shown on the diagrams on pages 59 through 61.

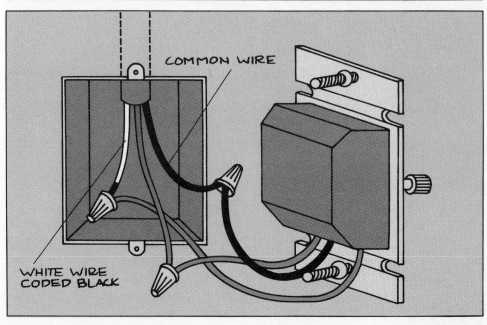

COMMON WIRE

WHITE WIRE CODED BLACK

Wiring multiple-switch circuits

The diagrams on the next three pages show how to wire Single-Pole, Double-Throw (SPDT) switches and special Double-Pole, Double-Throw (DPDT) switches. Both types enable you to turn fixtures on and off from two or more locations. DPDT switches are sometimes called four-way switches because they have four terminals. They can be combined with SPDT switches to control a light or outlet from *three or more* locations. The switches have only two positions however.

Controlling fixtures from more than one location is desirable in many situations. For example, stairways and long hallways should have light control at both ends. Control of garage lights from several locations is often useful. Outlets that can be turned on and off from several places are also handy for TV sets and music systems.

The following diagrams can help you add switches to an existing fixture or outlet (perhaps already having single-switch control) or plan entirely new installations.

To pick the diagram best suited to your project, decide first where the power-source cable will be in the circuit. If you are modifying an existing outlet or fixture, check the fixture, outlet, or switch boxes to find the source cable. Page 51 tells you how to identify the cables in the boxes.

NOTE:

SPDT—denotes a single-pole double-throw switch

DPDT—denotes a double-pole double-throw switch

A white wire coded black is represented as follows:

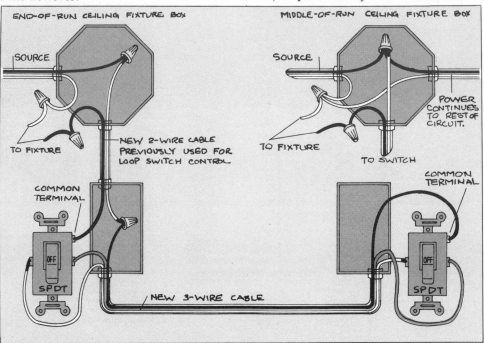

Two-switch control of ceiling fixture (Source power available at fixture box)

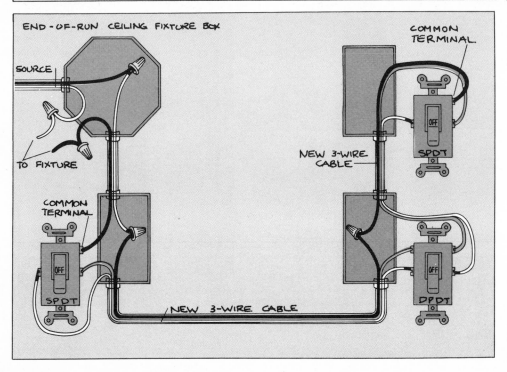

Three-switch control of ceiling fixture (Source power available at switch box)

What it takes

Time: 2 to 4 hours.

Tools and materials: Two single-pole, double-throw (3-way) switches; one or more double-pole, double-throw switches (4-way); two-wire and three-wire cable; screwdriver; wire stripper or knife; solderless connectors.

NOTE:

SPDT—denotes a single-pole double-throw switch

DPDT—denotes a double-pole double-throw switch

A white wire coded black is represented as follows:

Connecting new source cable

If the complete circuit is new, the source cable, shown entering one of the boxes, must be connected to another source. By using the appropriate diagram, you can join the power-source cable to this circuit at the fixture, outlet or switch box, whichever is most convenient. See pages 52 and 53 for diagrams showing how to connect new-circuit source cables to existing wiring at various points.

Grounding cable

The diagrams below apply for either steel-armored or plastic-sheathed cable. For simplicity, ground wires are not shown. However, ground wires must be connected. If plastic-sheathed cable is used, the ground wires should be joined and connected to the boxes, as shown on pages 52 and 53. If steel-armored cable is used, the ground wires should be connected to the cable where the cable enters a box.

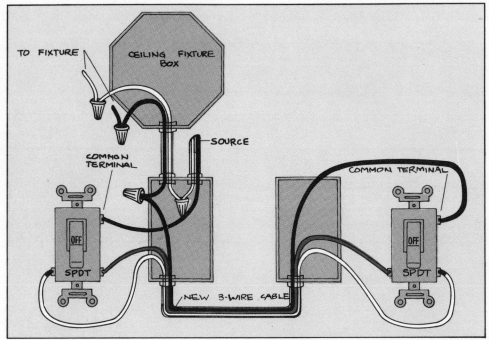

Two-switch control of ceiling fixture (Source power available at switch box)

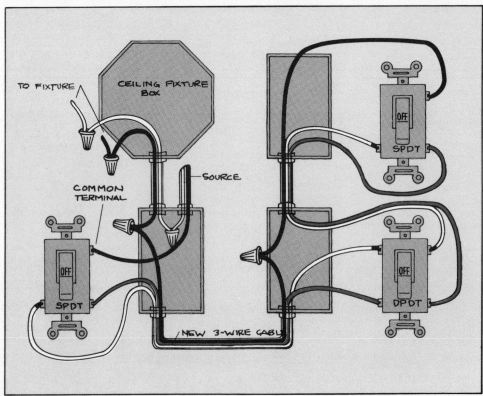

Three-switch control of ceiling fixture (Source power available at fixture box)

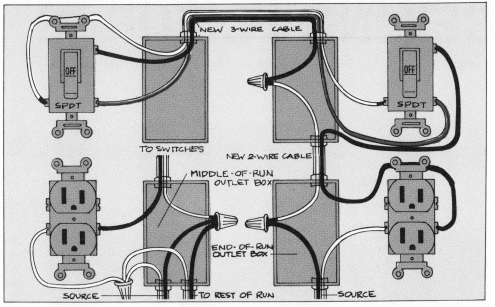

Two-switch control of wall outlet (Source power available at outlet box)

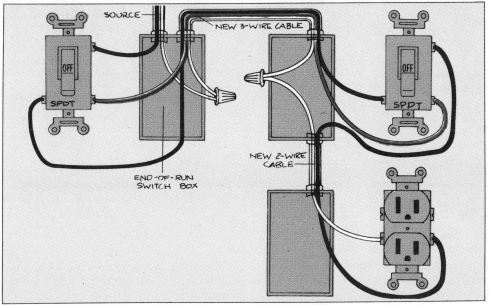

Two-switch control of wall outlet (Source power available at end-of-run switch box)

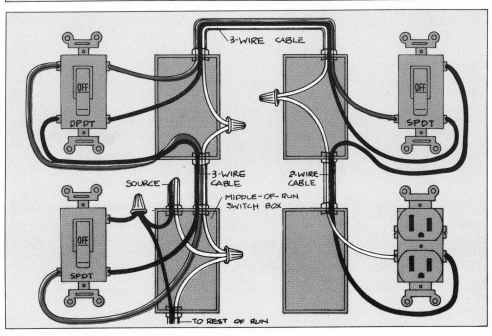

Three-switch control of wall outlet (Source power available at middle-of-run switch box)

6. CEILING FIXTURES

How to install a ceiling box

What it takes

Approximate time: 2-4 hours.

Tools and materials: Power drill; ⅛-inch bit; keyhole saw; carpenter's square; patching plaster; ceiling box and hanging hardware (for plaster-and-lath ceiling, a hammer and chisel); screwdriver; steel rule; 12- to 16-inch piece of stiff wire; masking tape.

The procedure to be followed when installing a ceiling fixture depends, first of all, on whether or not the ceiling is accessible from an attic or crawl space above it.

If the ceiling is accessible from above, the procedure is fairly simple and can be done with little or no visible damage to ti.e ceiling. Cable routing is also easier.

When the ceiling is not accessible from above (that is, when it is between finished floors), the fixture must be installed from below. This requires cutting into the ceiling and patching the opening when the job is complete. Routing the new cable to the ceiling fixture is also more difficult when working between floors.

Planning a switch?

A key factor in your choice of location for a ceiling fixture will be the type of switch to be used and its location. The tips on switch wiring below include how to connect a ceiling fixture with a built-in switch. For how to wire a wall switch to a ceiling fixture see diagrams on page 65. For details on cable routing see pages 66–68.

- Whenever practical, select a ceiling fixture with a built-in switch. This eliminates the wiring, cable routing, and carpentry involved in adding a wall switch.
- Switched fixtures are available in a wide variety of styles, ranging from simple porcelain pull-chain fixtures suitable for a garage or utility room to elaborate, counterweighted chandeliers suitable for living and recreation areas.
- To install switched fixtures, you need only connect the ceiling box to a power source. If the area above the ceiling is not finished, it is often possible to obtain power for the new fixture from an existing ceiling box in the same area. Check the section on ceiling-box wiring to determine the right way to make the connection.
- Once you have found a power source for the new ceiling box and have completed the box and wiring installation, you need only connect the black and white power wires to the black and white fixture wires. Use solderless connectors and connect black to black and white to white. If the fixture has screw-type terminals, connect the black wire to the brass screw and the white wire to the chrome screw.
- There are many situations, of course, when single or multiple wall-switch control of a ceiling fixture is desirable for appearance, convenience, or safety. For these installations some additional planning is necessary.
- First check existing wiring to find the most convenient place to connect your new fixture-and-switch circuit to a power source. Be sure to consider various switch positions. Sometimes routing the new cable can be made much easier if you investigate all possible switch locations.

When ceiling is accessible from above

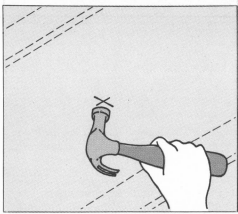

1. Check the area above the ceiling to determine the direction in which the joists run. Working from below, test the area where the fixture is to be located to be sure it is clear of joists and bridging. See detailed instructions at right. Locate the fixture at least 4 inches from a joist.

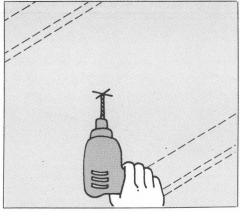

2. Drill a small (⅛-inch) pilot hole in the ceiling at the point you marked. This hole will locate the center of the opening you will make in the ceiling. If the area over the ceiling has flooring, you will need an extension bit so that the pilot hole will be visible from the room above.

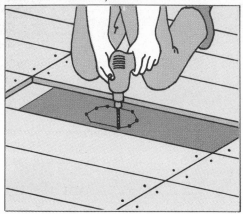

3. In the area above the ceiling, find the pilot hole. If necessary, remove a floorboard. Use the new ceiling box as a template. Center it over the pilot hole and trace the outline. Drill a ⅜-inch hole at each corner of the outline.

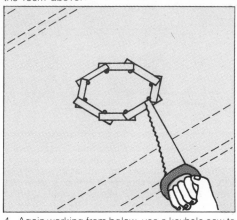

4. Again working from below, use a keyhole saw to cut through each section between the holes. If the ceiling is plaster and lath, use tape around the outline of the box to prevent cracking and chipping the adjacent plaster. Saw carefully and slowly. Protective goggles should be worn.

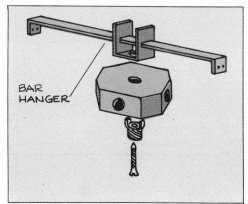

BAR HANGER

5. When the box cutout is finished, work from above again. Loosely connect the ceiling box to the bar hanger. Bend back or break off the tabs at the ends of the hanger so that the hanger will fit flush against the joists. If the best location for the ceiling fixture should turn out to be within an inch of a joist, a side-mounting ceiling box can be used. Hold the box against the joist and mark the mounting-hole locations. Prepare the cable and attach it to the box as described at right. Drill pilot holes in the joist at the points marked. Use wood screws to mount the box to the joist.

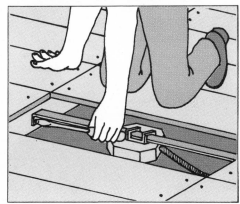

6. Prepare the end of the fixture cable and install it in the box. Secure it with a clamp. (See pages 20-25.) Hold the hanger and box so that the edge of the box is even with the lower surface of the ceiling. Spread the hanger to fit the space between the joists. Tighten the stud screw to secure the box on the hanger. Holding the box and hanger in place, mark the joist with the positions of the screw holes at the ends of the bar hanger. Drill the pilot holes in joists. Mount the box and bar hanger.

Checking for clearance:

To be sure the area above the fixture site is between joists and clear of bridging, make a small (⅜-inch) test hole in the ceiling. Bend a piece of stiff wire into a right angle 4 inches from the end and insert this end into the hole. Rotate the wire to feel for obstructions. If there is not sufficient clearance, drill another hole several inches away and test again.

When ceiling is between finished floors
With plasterboard ceiling

There's my way and the easy way

It seemed logical to run the new cable straight up the wall from the wall box to the ceiling, then directly across to the center of the ceiling for my overhead fixture. It was logical, but it was also dumb. I didn't realize that the second-floor joists my ceiling was attached to ran *at right angles* to the route of my cable. I would have had to cut a channel through the ceiling plaster and notch every dang one of the joists to fit that new cable flush.

Moral: Don't try to run cable across the ceiling from east to west, if your joists run north and south.

Practical Pete

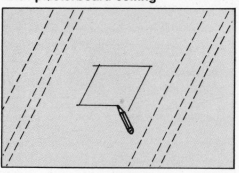

1. Make certain the area above where you plan to install the fixture is between joists and clear of bridging. You will need at least 4 inches of clearance in all directions around the fixture installation point. See detailed instructions on page 63.

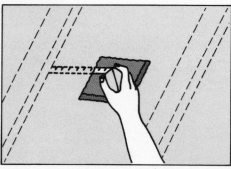

2. Cut a hole around the test hole, large enough for your hand to go through comfortably. Hold a steel rule in the space above the hole and measure the distance from the edge of the hole to each joist. Draw a line on the ceiling corresponding to the edge of each joist.

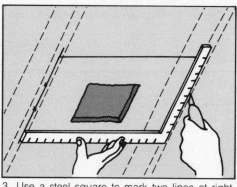

3. Use a steel square to mark two lines at right angles to the joist lines. The space between these lines should be sufficient (8–12 inches) to allow you to mount the ceiling box and hanger between the joists when you have cut out the rectangle of plasterboard. Drill a ⅜-inch hole at each corner of the rectangle. Use a keyhole saw to cut out the rectangle of plasterboard. A saber saw with a slow speed-setting can also be used, but be prepared for considerable plaster dust.

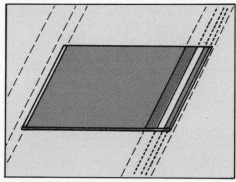

4. Cut two 1- by 2-inch furring strips slightly longer than the sides of the opening parallel to the joists. Secure the furring strips to the joists with wood screws. Position each strip so that the bottom edge is even with the bottom edge of the joist. The piece of plasterboard that will close the opening will be nailed to these strips. Cut a section of plasterboard to the dimensions of the opening. Locate the ceiling-box position on this piece of plasterboard. Draw the outline of the ceiling box, using the box as a template.

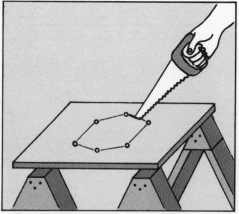

5. Drill a ⅜-inch hole at each corner of the outline and cut out the box opening with a keyhole saw. Loosely connect the ceiling box to the hanger. Bend back or break off the tabs at the ends of the hanger so that it will fit flush against the joists. Prepare the end of the fixture cable by removing the outer covering and conductor insulation (see page 20). Remove one of the box knockouts and install the cable in the box. Secure the cable with a clamp suitable for the type of cable you are using. (See pages 24-25).

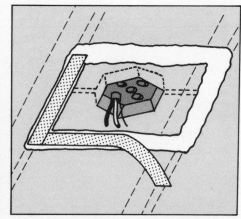

6. Hold the hanger and box in place so that the edge of the ceiling box is even with the surface of the ceiling. Spread the hanger to fit the space between the joists. Tighten the stud screw to secure the box. Again hold the box and hanger in place and check the position. Mark the joists with the positions of the screw holes, and drill pilot holes in the joists. Use wood screws to mount the box and bar hanger. Nail the plasterboard square to the furring strips to close the opening. Fill the cracks around the square with patching material and sand it smooth.

With plaster-and-lath ceiling

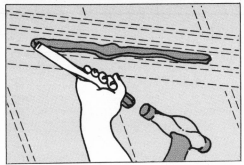

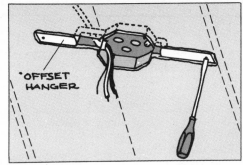

1. Make certain the area above where you plan to install the fixture is between joists and clear of bridging (see page 63). When a suitable location has been found, cut a hole around the test hole large enough for your hand to go through comfortably. Hold a steel rule in the space above the hole and measure the distance from the edge of the hole to each joist. Put a mark on the ceiling corresponding to the edge of each joist. Use a hammer and chisel to chip out a channel in the plaster between the joists. The channel should be wide enough and long enough to accommodate the offset hanger. Widen the channel at the point where the box will be mounted. Use the box as a template to mark the plaster so that the box opening will be large enough.

2. Assemble the ceiling box to the offset hanger. Prepare the end of the fixture cable by removing the outer covering and conductor insulation (see page 20). Remove one of the box knockouts, and install the cable in the box. Secure the cable with a clamp suitable for the type of cable you are using. (See pages 24–25.) Secure the offset hanger to the joists with wood screws. Fill the channel with patching plaster, but leave room for a finishing coat of spackling compound or plasterboard joint cement. Allow the patching plaster to dry thoroughly before applying the finishing coat. Sand the finish coat smooth when the plaster has dried.

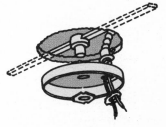

A special hanger is available for mounting ceiling boxes between finished floors. This mount requires only a small ceiling hole. The bar rests on the plasterboard or plaster-and-lath ceiling material. A shallow box mounts on the bar. The space available in the box is limited but adequate for connecting power-cable conductors to a fixture. Because the bar rests on the ceiling material and is not fastened to the joists, this hanger should be used only for fixtures under 5 pounds.

Wiring a ceiling box to a wall switch

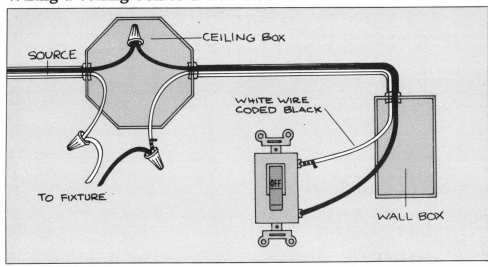

Source power connected at ceiling box

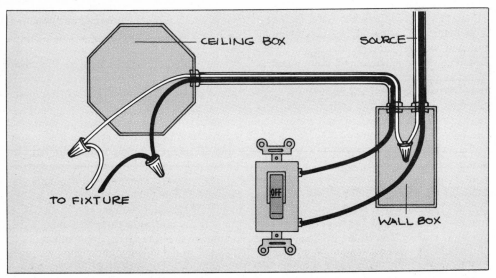

Source power connected at wall box

Running cable to ceiling box

If the wall and ceiling material is plasterboard or conventional plaster and lath, the wall material is not thick enough to cover the cable at the wall-ceiling opening. The top plate must be notched to recess the cable. Use a saber saw to make vertical cuts in the top plate, about 1 inch apart and ½ inch deep. Use a chisel to chip out the wood between the cuts. Push the cable into the notch and secure it with a cable staple.

JOIST

CEILING-PLATE HOLE

WALL-PLATE HOLE

STUD

What it takes

Approximate time: This involves carpentry and planning. Allow more time than you think you will need and remember that the power must stay off on all the circuits you are working with until you are done, so work in the daytime. Plan on a good eight hours for each between-floor run.

Tools: Hammer, utility knife, nail set, keyhole or saber saw with fine-toothed blade, electric hand drill, 1/16-inch twist-metal bit and ¾-inch spade bit with 18-inch drill-extender attachment, fish tape, electrician's tape, screwdriver, chisel (to loosen baseboards or notch studs), multipurpose tool for stripping insulation off conductors, voltage tester, safety goggles (if you must install a box in a plaster ceiling).

Materials: Solderless connectors, nails, cable, fixtures, junction boxes, wall boxes, receptacles, switches, and accessory hardware required in your wiring plan.

Between finished floors

It takes a bit of work to run cables from walls to ceiling when the ceiling is below a finished floor.

Access holes must be cut in the wall and at the point where wall and ceiling meet. (These are patched when the job is done.) Top plates and sometimes studs must be notched to accept the cable.

In this example the power connection is being made at a wall receptacle. As shown, the cable runs laterally across two studs, up through the wall to the switch, from the switch to the ceiling, and between joists to the ceiling-box opening.

To start, you would cut access holes where the wall and ceiling meet and at each stud. Through the access holes, notch the studs so you can run the cable laterally across them. Use a saber saw to make two horizontal cuts, about 1 inch apart, in each stud. The cuts should be ½ inch deep. Chisel out the wood between the cuts.

The access hole at the point where wall and ceiling meet should be large enough to extend below the wall top-plate and into the ceiling itself.

The best way to make the holes in the wall and ceiling depends on the material. For plasterboard, the hole can be cut with a utility knife, or a pilot hole can be drilled and the material cut out with a keyhole saw. Plaster walls with wood lath can be similarly cut. If the plaster is over metal lath, a saber saw with a fine-toothed cutting blade can be used.

When you have cut the top-plate hole and the corresponding lower access hole, you can select a spot for mounting the switch. The switch box can be mounted on a stud or between studs. The various methods for mounting boxes are described on pages 63–65. Whichever method you choose, use the switch box as a template. Mark the outline on the wall. Drill a pilot hole and cut the opening. Do not mount the box until the cable has been "fished" through.

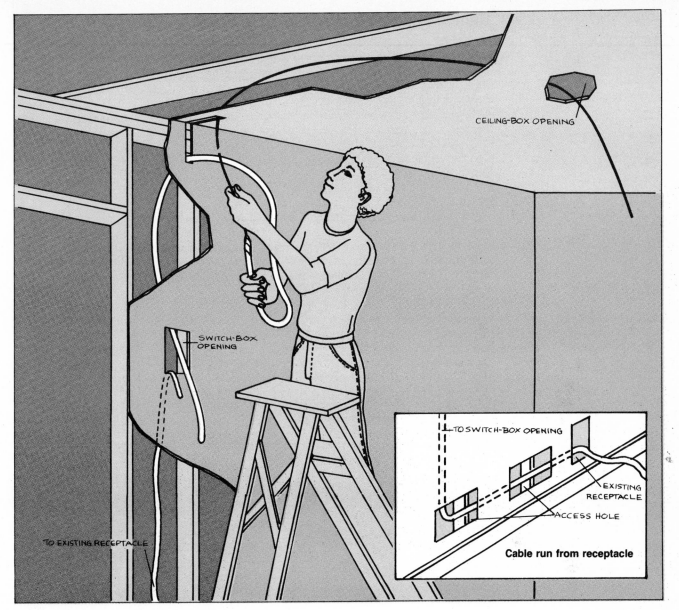

CEILING-BOX OPENING

SWITCH-BOX OPENING

TO EXISTING RECEPTACLE

TO SWITCH-BOX OPENING

EXISTING RECEPTACLE

ACCESS HOLE

Cable run from receptacle

Starting at the ceiling-box opening, feed the cable into the hole in the direction of the ceiling hole at the top plate. Use a fish tape or a piece of wire to locate and pull the wire down through the ceiling hole. Next, feed the cable into the wall through the hole below the top plate. Use the fish tape or wire again to pull the cable out through the switch-box opening.

Continue to pull the cable through the switch-box opening, while a helper feeds it into the ceiling-box opening, until there is about a foot of cable extending through the ceiling box.

Cut the cable extending through the switch-box opening. Again, leave a foot or so of cable hanging out. Power must be turned off at the circuit-breaker or fuse panel so there is no power at the receptacle box before you do the next step. *Test before you touch.* At the receptacle box where the power connection is to be made, remove one of the box knockouts and feed a length of cable through the knockout hole in the direction of the first stud-access hole. Use a fish tape or a piece of wire to pull the cable through the wall and around the first stud. Repeat this procedure until the cable is routed around each stud.

At the end of the lateral run, feed the fish tape down from the switch-box opening to the bottom access hole. Connect the fish tape to the cable that has just been run laterally from the receptacle box. Pull the cable up and out of the switch box.

Tag or mark the two cable ends at the switch opening so you know which is the source cable and which end goes to the ceiling fixture. Feed the cables through knockouts in the new wall box. Mount the box on the stud or on the wallboard. (See pages 63–65.) Use the internal cable clamp in the box to secure the cables.

Make the electrical connections for this type of installation as shown on page 65 ("Source power connected at the wall box").

How to scare a cat
Go easy with the hammer when you use staples to attach plastic-sheathed cable to studs. Drive the staple into the wood only far enough to hold the cable firmly. If you drive it in too far, the staple can damage the sheathing—even short the conductors. Even if all the damage you do is to scare eight lives out of your cat, you'll still have to splice or rerun the cable.

Practical Pete

Ceiling-box cable run in attic or crawl space

In an attic or crawl space, cable runs are simple and straightforward. If the area contains flooring, you will have to remove some boards temporarily and route the cable through or along the joists.

If you need to run a cable from the attic to the basement, perhaps for power tie-in, check the space around the plumbing vent pipe. It usually provides a clear, open area from attic to basement.

To run cable from the attic to walls for switch connections, it is only necessary to drill through the wall top plate. The top plate consists of two 2-by-4s, sometimes topped with a 1-by-4. This means you must drill through approximately 4 inches of lumber. A long bit or a paddle bit must be

used. If you are using a light-duty drill, stop occasionally to allow the drill to cool.

When making measurements in an attic or crawl space, always use a reference point that is visible on the floor below. For example, measure from the edge of a stairway opening or the access-hole opening to the point on the wall where you plan to mount the switch. If necessary, measure in two directions at right angles. The idea is to make measurements that can be repeated with reasonable accuracy in the attic or crawl space. Remember, it's easy to misjudge locations when you are working in an unfamiliar area. Careful measurement and checking will save unnecessary work and perhaps an unexpected patch job.

Short tempered

It is tough on the temper to discover that the cable you just fished through the wall is almost—but not quite—long enough. The way to avoid this is simple—cut long.

A good rule is to add about 20% to your straight line measurement. Cable does not bend sharply, lie perfectly flat, or even hang straight. You may also find you have to go around an unexpected obstruction.

Practical Pete

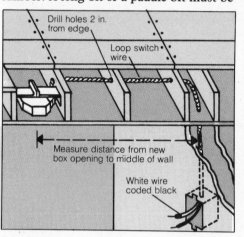

Remove one or more floorboards and drill joists as shown to run cable from top plate to ceiling box in a floored area.

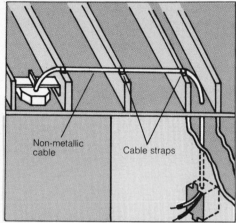

If area has no flooring, simply run cable across joists and secure with cable straps.

To route new cable to existing wall box

1. Drill a hole in the top plate just above the box.
2. Then remove a knockout from the top of the box.
3. Have a helper drop a weighted string through the top-plate opening.
4. Use a fish tape or a piece of wire to hook the string and pull it through the box opening. Use a weight that will pass through the box opening (for example, a medium-size bolt). If you cannot locate a suitable weight, use any disposable piece of metal and simply cut the string after you have fished a loop through the hole.
5. Attach the cable securely to the string.
6. Have your helper pull the string up slowly while you feed the string and cable through the box opening and top-plate opening. The cable will go through openings more readily if you smooth the connection between cable and string by wrapping it with tape.

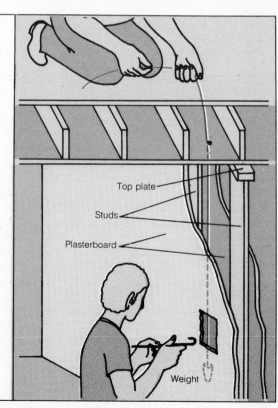

Mounting a ceiling fixture

What it takes

Approximate time: Half an hour. More if you must wire and install a wall or ceiling box for the fixture.

Tools and materials: Screwdriver, utility knife, pliers, crescent wrench to loosen stuck nuts on old, large fixtures (use pliers to tighten nuts on new fixtures, not a wrench), tape, solderless connectors, wire coat hanger bent in S-shape, voltage tester.

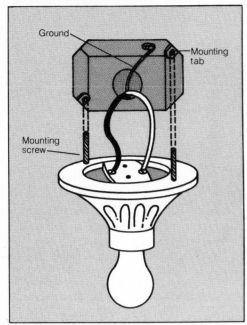

Directly to the ceiling box. A simple, porcelain single-bulb fixture can be mounted directly to the ceiling box with two screws. These fixtures have screw-type electrical connections. Attach the black wire to the brass-colored terminal and attach the white wire to the silver-colored terminal.

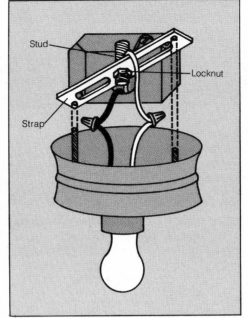

Stud mounting. If the box has a fixture stud, the strap can be secured to the stud with a locknut. The fixture is then mounted on the strap with screws.

Electrical connections are made by means of solderless connectors. Connect the black power wire to the black fixture wire and connect the white power wire to the white fixture wire.

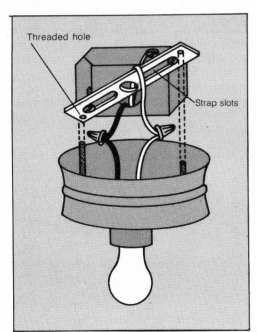

Strap mounting. Mounting a strap on a ceiling box is done by putting two screws through the strap slots and screwing them to the threaded ears in the outlet box. The fixture is screw-mounted to the holes in the ends of the strap. The slots in the strap allow you to shift the fixture position so that it need not be directly centered under the box.

Electrical connections are made by means of solderless connectors. Connect the black power wire to the black fixture wire and connect the white power wire to the white fixture wire.

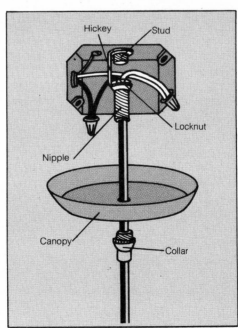

Large fixture mounting. To mount larger and heavier fixtures, first thread a hickey onto the fixture stud and then secure a nipple to the hickey with a locknut. The nipple extends through the base of the fixture. Thread the collar on the fixture onto the nipple to hold the fixture canopy snugly to the ceiling above it.

Electrical connections are made by means of solderless connectors. Connect the black power wire to the black fixture wire and connect the white power wire to the white fixture wire.

7.FLUORESCENTS

Fluorescent lighting provides a bright, even illumination that is desirable in many areas of the home. It is more complex and, initially, more costly than incandescent lighting. Fluorescent lamps, however, produce more light per watt of power used, and they last four or five times longer than incandescent lamps. The long-range cost of fluorescent lights, then, is significantly lower than the cost of incandescents.

The life of fluorescent lamps is determined primarily by how often they are turned on and off; the less this occurs, the longer they last. Since fluorescents use little power, it is better to leave them on than to turn them on and off for short periods.

The three most common types of fluorescent fixtures in homes

Rapid start. This is currently the most widely used type of fixture. It lights almost immediately when switched on. Rapid-start fixtures also have the advantage of being readily adapted for use with fluorescent dimmer switches. Lamps for rapid-start fixtures have two pin-type connectors at each end.

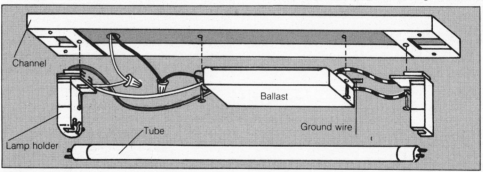

Instant start. This type of fluorescent fixture lights a second or two after it is switched on. It requires a higher initial voltage surge than other types. Lamps for instant-start fixtures may have one or two pin-type connectors at each end. On this type of fixture, the lamp holder contains a built-in switch that allows high voltage to be applied only when the fixture contains a lamp.

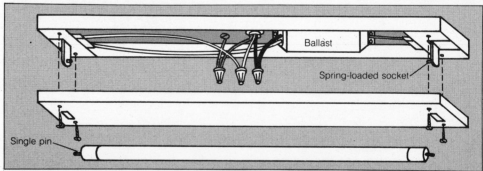

Starter type. This type of fixture has a separate starter. It uses lamps that have two pin-type connectors at each end. The starter, like the lamps, has a specified life, but is replaceable. It is located in a socket near one of the lamp holders. For replacement, power to the fixture is turned off, and the lamp is removed. Then the starter is twisted and pulled out of its socket. Replacement starters must match the wattage of the lamp.

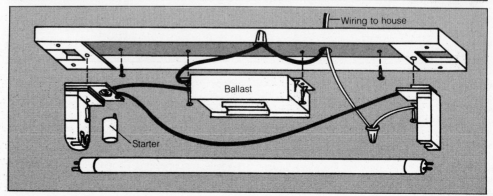

Fluorescent characteristics

Fluorescent light and incandescent light have different characters, and, generally, the two do not blend well. Some recently developed fluorescent lamps, however, produce illumination that is compatible with incandescent lighting. These include the warm-white types and those that have the word *deluxe* in the type name. Warm-white and deluxe-type lamps can be effectively combined with incandescent lights. It is also possible to add dimmers to fluorescent fixtures to control the light intensity. This, too, is a help in blending fluorescent and incandescent light. When blending, never use more than one type of fluorescent lamp, as different types often do not blend well and create a poor mix of lighting.

For areas where maximum light is important—workbenches, kitchen work surfaces, and desks, for example—use cool-white and daylight-white lamps. They produce the maximum light per watt of power. For decorative use in living areas, warm-white and deluxe types are better, as they produce a softer light. Warm-white and deluxe-type lamps put out about 30 percent less light per watt of power than do cool-white lamps.

This chart summarizes the most commonly used fluorescent lamps.

Color characteristics of commonly used fluorescent lamps

Lamp description	Atmosphere	Effect on colors	Compatibility with incandescent light
Cool white (standard)	Cool	Intensifies cool colors; dulls warm ones. Light output high.	Poor
Cool white (deluxe)	Cool	Improves appearance of all colors. Light output medium.	Poor
White	Warm	Little effect on cool colors; dulls warm colors slightly. Light output high.	Fair
Warm white	Warm	Distorts all colors slightly. Light output high.	Good
Warm white (deluxe)	Warm	Leaves colors true. Flattering to human complexions. Light output medium.	Very good
Living white (natural)	Cool	Especially designed to flatter human complexions. Leaves all colors reasonably true. Light output medium.	Very good
Cool green	Cool	Intensifies cool colors. Light output high.	Poor

How fluorescent lights operate

Fluorescent lights and incandescent lights operate on entirely different principles. All fluorescent lamps contain a small filament (similar to the filament of an incandescent lamp) at each end. The glass tube is filled with a gas (mercury vapor) and the inner surface of the tube is coated with a phosphorescent substance.

When the current is turned on, power is applied to the filaments, causing them to heat up. The hot filaments vaporize the gas in the tube, making it a good conductor of electricity. The filaments are then turned off and a high-voltage surge of power is momentarily applied to the tube. The surge starts current flowing through the tube. Once the current flow is established, it continues with only normal line-voltage applied. In fact, current flows so easily in the vaporized gas that it must be limited by a device called a ballast.

The flow of current through the gas produces ultraviolet light. Although it is barely visible to human eyes, the ultraviolet light causes the phosphorescent coating on the tube to emit strong and visible light.

The ballast is a special type of transformer that produces the high-voltage surge necessary to start current flowing in the fluorescent tube. Once the current flow has been established, the ballast limits the flow through the tube to the rated value. The limiting action is required because, once current is flowing in a fluorescent lamp, the lamp's internal resistance drops to a low value. If not limited, the current would destroy the lamp in a short time. The ballast prevents this.

The starter found in some older fluorescent fixtures is a small metal canister that fits in a socket on the fixture. The starter does the switching needed to turn the filaments on and off, apply the high-voltage surge to start the lamp, and switch in the ballast to limit the current.

Installation in halls, kitchens, large areas

What it takes

Approximate time: If you need no carpentry or plastering work, you can mount a new fluorescent fixture onto an existing outlet or ceiling box wired for an old incandescent fixture in about an hour.

Tools and materials: Hammer, screwdriver, pliers, voltage tester, solderless connectors.

It is practical to replace incandescent fixtures in your home with fluorescent ones anywhere that they will not be turned on and off frequently. The channels of fluorescent fixtures are provided with a number of knockouts, any one of which can be used for mounting.

Remove the channel cover, tap the knockout with a hammer to break a section free, then grip the knockout with pliers and twist it off.

Next, turn off the power to the incandescent fixture and remove it. Mount the fluorescent fixture by placing the knockout over a threaded nipple and securing the

fixture with a washer and locknut.

(You can add a strap-mounted nipple to wall outlets or a hickey and nipple to ceiling boxes that do not contain nipples.)

Feed the black and white fluorescent power-wires through the nipple and join them to the source-cable conductors with solderless connectors. Connections are black-to-black and white-to-white.

Large fixtures have a mounting cutout. To mount these fixtures, place a metal strap inside the channel, across the cutout. Then place the fixture and strap over the nipple. Use a locknut to hold the strap and fixture against the ceiling.

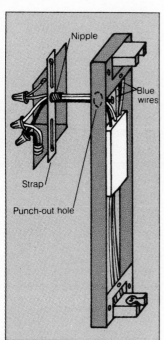

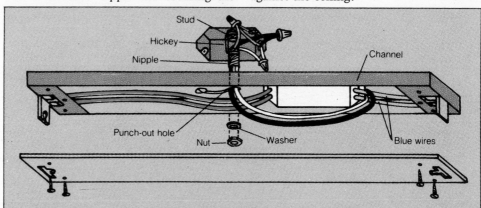

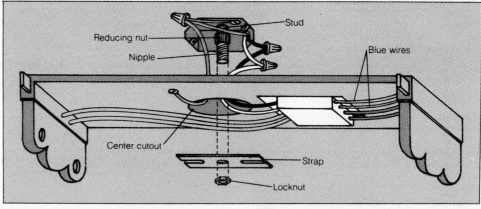

Circular fluorescent fixtures make good replacements for kitchen-ceiling fixtures. To make the substitution, first turn off power to the kitchen-ceiling box. Remove the cap nut or mounting screws and lower the incandescent fixture. Support the fixture so it does not hang by the electrical connections. Remove the solderless connectors that join the black and white fixture wires to the power wires. *Test before you touch.* Untwist the black and white fixture wires and set the incandescent fixture aside. Using solderless connectors, connect the fluorescent fixture wires to the power wires in the box (black-to-black and white-to-white is the invariable rule).

If necessary, use a hickey or reducing nut to install a nipple long enough to project through the fixture. Fold the wires into the fixture canopy and secure to the ceiling by tightening the cap nut on the nipple.

The simplest installation of all can be used for workbench lighting. Purchase a complete fixture. Twin-lamp, 48-inch fixtures, complete with reflector and switch, are widely available. Hooks and chains are supplied with the fixture. Attach the hooks to the ceiling and hang the fixture from the chain. To make the electrical connection, permanently wire the fixture to a nearby ceiling or wall box. If a spare grounded outlet is available, plug the fixture in.

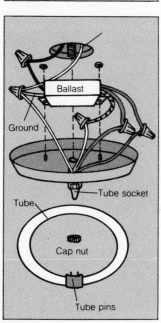

Installation in bathrooms

Fluorescent illumination is well suited to bathroom use—applying makeup, shaving, and so forth. The fixtures can be mounted on either side of, or above, bathroom cabinets; cabinets with built-in fixtures are also available.

Bathroom cabinets are generally recessed into walls and secured to studs with wood screws. To remove a cabinet, simply remove the screws on each side and lift the cabinet from the wall opening. The opening provides easy access to electrical connections on the interior wall. If an incandescent fixture was mounted near the cabinet, and is to be removed, the fluorescent fixture can be wired to the same box.

Turn off power to the wall box by removing the fuse or turning off the circuit breaker for that circuit. Connect the fluorescent-fixture wires to the same wires that were connected to the old fixture (black-to-black and white-to-white).

If the incandescent fixture was controlled by a wall switch, the fluorescent fixture will now be controlled by the same switch.

If you wish to retain the incandescent fixture with switch control and have separate control over the fluorescent light, two wiring schemes are possible. Switch and fixture wiring in bathrooms is the same as switch and ceiling-fixture wiring (page 52).

Inspect the wall-box wiring and compare it with the switch and ceiling-fixture diagrams on page 52 to determine if source power is available at the wall box. If it is, you can wire the fluorescent fixture to the source-power cable and turn it on and off by means of an individual fixture switch.

If a new circuit is required for the fluorescent lights, install a junction box in the wall stud behind the cabinet.

Power for a junction box is obtained by running a cable to an existing wall or ceiling box. Connect the leads from the fluorescent fixtures to the power cable in the junction box with solderless connectors. **A word of caution:** Many local codes now require that GFIs (ground fault interrupters) be installed on bathroom circuits. If there is one on the bathroom circuit in your home, be sure you make the power connection for the junction box to a wall or ceiling box on the GFI-protected circuit. You must, of course, turn off power to the existing wall or ceiling box before you connect the new cable. Make sure that when the bathroom-circuit GFI is turned off, power is off in the box you are using for the new circuit.

Color me green
My wife wanted fluorescents on each side of the bathroom mirror so she would have plenty of light to make up her face. So I put them in (she needs all the help she can get). The lamps I picked out had high light output and a refreshing, cool color. They were called "cool green." They made her look like the victim in a Dracula movie. Heck, anyone can make a mistake. If you don't want to make that one, see the chart on fluorescent lamp colors on page 71.

Practical Pete

Remove bathroom medicine cabinet for easy access to the inside of the surrounding wall when you want to add new, or replace old, wiring.

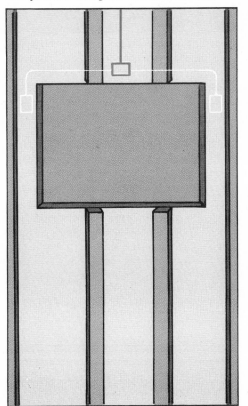

Rear view of wall with cabinet removed showing location of original wiring for incandescent fixture (red), and new wiring for vertically mounted fluorescent fixtures (white).

Fluorescent dimmers

With fluorescent dimmers, you can have full-range intensity control over fluorescent fixtures. This provides variety in lighting and is a help in blending fluorescent and incandescent light for more pleasing effects.

Fluorescent-dimmer circuits are really special lighting systems—that is, all the devices involved must be matched. The dimmer control, the fluorescent fixture, the fixture ballast, the lamps, and the wiring must be selected according to the number of lamps to be controlled.

For applications (such as cornice and valance lighting) where a maximum of eight 40-watt, single-lamp fixtures will be used, a two-wire dimmer system can be installed. For this, the existing wiring is sufficient.

For larger applications—such as luminous ceilings and walls—where more than eight single-lamp fixtures are required, a three-wire dimmer system is needed. For example, a luminous ceiling in a 12-by-15-foot room would require fifteen to eighteen 40-watt fluorescent lamps and, therefore, a three-wire system would be needed. The three-wire system can handle as many as forty 40-watt fluorescent lamps in either single or double fixtures.

For both the three-wire and two-wire dimmer systems, special ballasts must be installed in each fixture to be controlled. Rapid-start fixtures and lamps should be used in both types of dimmer systems.

What it takes

Approximate time: Half an hour.

Tools and materials: Voltage tester, small and medium screwdrivers, pliers, solderless connectors, utility knife, special dimmer ballast that matches dimmer control, fixture(s), and lamps.

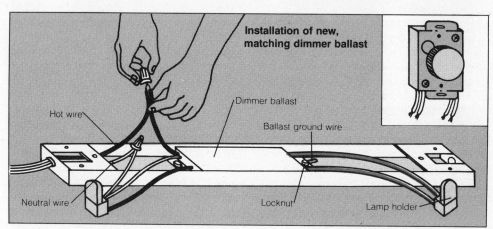

Installation of new, matching dimmer ballast

Hot wire • Dimmer ballast • Ballast ground wire • Neutral wire • Locknut • Lamp holder

Installing a two-wire-system dimmer ballast. To replace the ballast, disconnect the fixture from the power source wires for that circuit.

Remove the lamp, the screws that hold the channel cover, and then remove the cover. Next, remove the solderless connectors from the black and white wires. Untwist the wires. Remove the screws securing the lamp holders and those that hold the old ballast in place. Take out the ballast and lamp holders. Disconnect the lamp holders from the ballast wires. Some lamp holders are attached to the ballast wires by push-in type connectors. To release these connections, insert a small screwdriver in the slot next to the wire entry hole. Press in on the screwdriver and pull on the wire to release it. The short white wire in one lamp holder will be reused as is and need not be removed. Connect the dimmer ballast to the lamp holders at either end of the fixture as shown in the drawing.

The new connections will be the same as the original ones: two wires from each end of the ballast to each lamp holder. Mount the dimmer ballast and the lamp holders in the chan-nel. Make sure the bare ground wire from the ballast is wrapped securely around one of the mounting screws. Use the old solderless connectors to join the black power wire to the black ballast wire and the white power wire to the short white wire you will find on the lamp holder at one end.

The installation of a three-wire dimmer system for single-lamp fixtures is similar to the installation of the two-wire ballast, but two additional connections are required.

As shown in the three-wire system diagram, opposite, the ballast-control lead must be connected to the black wire in the three-wire cable.

Fixtures used on the three-wire system must have one circuit-interrupting lamp holder. This lamp holder has a built-in switch that is actuated by the pins on the fluorescent lamp.

If the fixture does not already have this type of holder, the holder wired to the blue and white ballast leads must be replaced with a circuit-interrupting type. Remove all three wires from the standard holder and connect the short white wire, as well as the blue and white ballast wires, to the circuit-interrupting holder, as shown opposite.

Dimmer controls are designed to fit in standard toggle-switch wall boxes. Boxes, however, may be mounted in various ways. So, if an existing box is to be used, check the amount of space in it against the dimmer-manufacturer's data. For new installations, use a box large enough to handle the dimmer and its connections conveniently. (You will have to make four connections, plus ground, in the wall box.)

If they are to operate properly, all elements of the dimmer circuit must be matched. All fixtures should be rapid start and of the same size and type. Four-foot, 40-watt, single- or double-lamp fixtures are usually specified, as are the recommended ballasts. All fixtures controlled by one dimmer must have the same type dimmer ballast. Nothing other than fluorescent fixtures should be on the dimmer circuit.

To get the most uniform illumination at all brightness settings, use only new lamps on a dimmer circuit. The new lamps should be operated at full brightness for the first 100 hours.

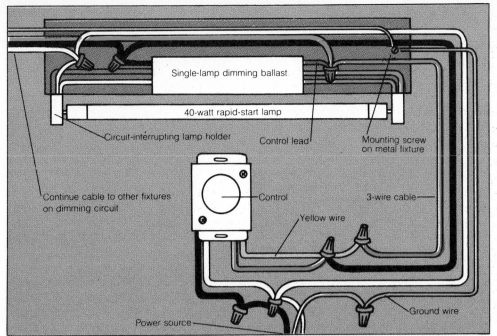

Single-lamp dimming ballast

40-watt rapid-start lamp

Circuit-interrupting lamp holder

Control lead

Mounting screw on metal fixture

Continue cable to other fixtures on dimming circuit

Control

3-wire cable

Yellow wire

Power source

Ground wire

Wiring diagram for rapid-start single-lamp dimmer ballast

The diagram at left shows how one fluorescent-dimmer system is wired. The system is intended for use with single-lamp fixtures. The number of single-lamp fixtures that can be added to the circuit is limited only by the rating of the control. Additional fixtures should be wired like the one shown. The ballast ground wire (not shown) must be attached to the fixture. To do this, wrap it around one of the mounting screws.

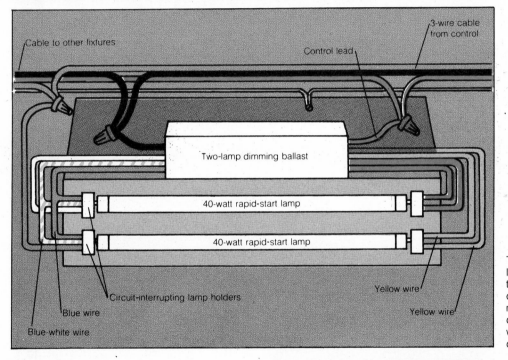

Cable to other fixtures

Control lead

3-wire cable from control

Two-lamp dimming ballast

40-watt rapid-start lamp

40-watt rapid-start lamp

Circuit-interrupting lamp holders

Blue wire

Blue-white wire

Yellow wire

Yellow wire

Wiring diagram for rapid-start, two-lamp dimmer ballast

This diagram shows how a two-lamp dimmer ballast is wired in a three-wire dimming system. The cable ground wire (not shown) must be attached to the fixture channel, and the ballast ground wire must be wrapped around one of the ballast mounting screws.

How to adjust fluorescent dimmers is described on the switch package or in a separate instruction sheet. For the most common types, the adjustment is made by means of an adjustment screw or a knurled collar on the switch-control shaft. Set the control shaft to the extreme counterclockwise position. Turn on circuit power. Turn the shaft to the extreme clockwise (brightest) position. With the shaft in this position, use a screwdriver to turn the adjustment screw (or pliers to turn the collar) clockwise as far as it will go. Now turn the control shaft to the lowest setting at which the fluorescents still light. Rotate the adjustment screw or collar counterclockwise until the lamps begin to flicker, then turn it back (clockwise) just enough to eliminate the flicker. With this adjustment, the lamp will provide light until the power is off.

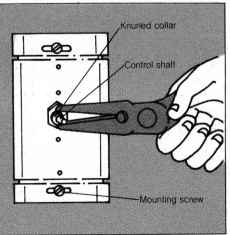

Knurled collar

Control shaft

Mounting screw

Joining fixtures

What it takes

Approximate time: Roughly an hour and a half, depending on the condition of previous fixtures and ceiling materials.

Tools and materials: Use a ⅝-inch bit and brace to make starter hole in ceiling. Insert keyhole saw in hole and saw away parts of ceiling material necessary to remove old fixtures. Other tools needed: multipurpose tool or wire stripper, cable clamps, solderless connectors, screwdriver, pliers, utility knife, hammer, nail set, voltage tester. Use type TW No. 14 wire for jumpers when wiring one fixture to another. Buy 5 feet of black and 5 feet of white wire for each 4-foot-long fixture. Use plastic-sheathed or armor-covered cable between fixtures, and to the power-source.

In many decorative fluorescent installations—such as valance and cornice lighting or luminous walls and ceilings—it is necessary to wire several fixtures from a single power source and to control them with a single switch. Run jumper wires from the power connections in one fixture through the channel to the next fixture, and so on. Two types of fixtures offer different ways of connecting channels.

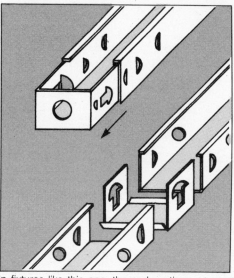

In fixtures like this one, the end sections are removable. After they are removed from adjacent ends, one of the end pieces can be used to join the channels together.

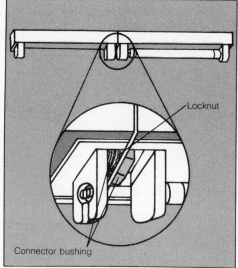

In this type of fixture, a knockout is removed from adjacent ends of the fixtures and a connector-bushing fitting, secured with locknuts, is installed to join the channels.

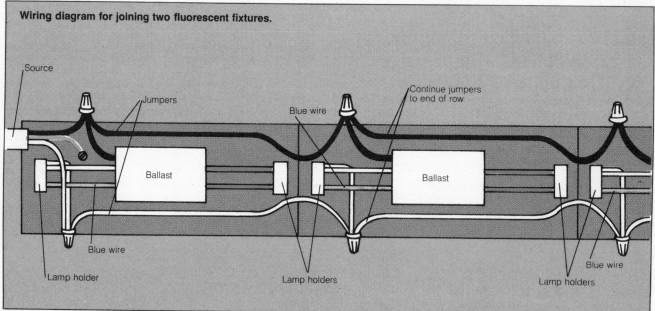

Wiring diagram for joining two fluorescent fixtures.

To wire one fixture to another you will need type TW No. 14 wire for jumpers. If you are joining 4-foot fluorescent fixtures (the longest available fixture is usually the easiest to use), figure on using 5 feet each of black and white wire per fixture. Use plastic-sheathed or armor-covered cable to connect one row of fixtures to another and to connect the fixtures to the source box. Type TW wire should be used only inside the fluorescent channels. Be sure to purchase wire with black and white insulation. The conductors in the wires are the same; but, if you maintain the black-to-black and white-to-white connection col-

ors all the way, you will be less likely to make a wiring mistake. Further, coded wires make future changes, additions, or maintenance easier. In the first fixture, join the black power wire, the black ballast wire, and a black jumper wire with a solderless connector. Join the white power wire, the short white wire from one lamp holder, and the white jumper wire with a solderless connector. Connect the bare ground wire from the power cable to any convenient channel screw. Lay the jumpers in the channel. Run one on each side of the ballast. Feed the jumpers through the connector bushing between channels, or

simply run them along the channel to the next fixture, according to the method used to connect the fixtures. Make similar black and white wire connections at the next fixture. Continue until all the fixtures are joined. The bare-wire ground connection made at the first fixture will ground all fixtures joined together metal-to-metal. If another row of fixtures is to be installed, run a ground jumper from the power-cable ground connection in the first row to the first fixture in the second row. Wrap the ground jumper around a ballast mounting screw Use green-insulated wire for the ground jumpers.

For multiple-row installations, such as those used to create luminous ceilings, connect ground jumpers to the power-cable ground wire at the source box. Connect as many jumpers as necessary to provide one direct ground connection for each row of fixtures. This assures you of a good ground for each row. Rapid-start fixture channels must be grounded to ensure proper operation. Where luminous ceilings are being installed, a ceiling box previously used for an incandescent fixture is often available. If the ceiling box contains a source cable, the rows of fluorescent fixtures can be wired to the box, as shown. If a loop circuit to a wall switch is also available at the ceiling box, the same switch circuit can be used to control the fluorescent fixtures.

Since the old ceiling material will be concealed by the new luminous-ceiling panels, you can simply cut away ceiling material around the box, as necessary; remove knockouts from the box; and install

cable clamps for the new wiring.

Another method of connecting the new cables is to add an extender to the ceiling box. The extender is simply another box with the back cut out. Use screws to attach the extender to the existing box. Remove knockouts from the extender, and install cable clamps for the new wiring. Whichever method you use, be sure to install a cover plate on the box or extender when the wiring has been completed.

When a ceiling box with source cable is not available, power for the luminous ceiling must be brought from another source—such as a wall outlet. In this case, the fixture rows can be wired as shown. The cable from the wall box to the ceiling need not go through the wall top plate if the new ceiling is to be below that level. The power cable can be brought through the wall at any convenient point. It is easiest to start wiring from an end row. Route the cable to the nearest corner and then wire as shown.

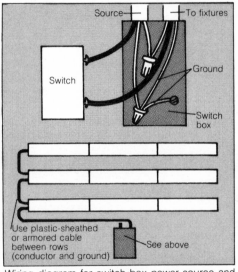

Wiring diagram for ceiling-box power source and cable run from box to rows of fluorescent fixtures.

Wiring diagram for switch-box power source and cable run from box to rows of fluorescent fixtures.

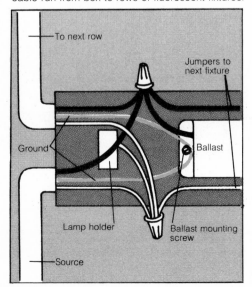

Wiring diagram for cable into fixture between rows of fixtures.

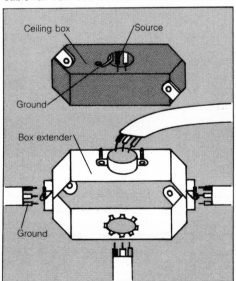

Wiring cables into a ceiling box extender.

Black and white and red all over
Sure, you *can* buy TW wire with conductors the same color, but don't. Make sure the insulation on the two wires is of different colors. When I started connecting all those wires inside the fixture, and running from one to the next, I went plumb batty trying to keep track of which was the "white" wire and which "black" so I wouldn't hook a neutral wire to a hot one and blow the whole shebang. Wiring diagrams are hard enough to follow without complicating the job even more.

Practical Pete

Two-circuit wiring

Luminous ceilings and walls can be wired so that some of the fixtures are on one switch circuit and the rest are on another circuit. This provides two levels of illumination without the work and expense of installing dimmer systems.

In a recreation room, for example, you may want low-level lighting for watching television and full illumination for card playing and similar activities. You can use the same scheme for valance, soffit, and cornice lighting. You can wire any number of fixtures, and fixtures of different sizes, on each circuit to provide whatever light levels you want.

In an unfinished room where you have easy access to wall studs you can use plastic sheathed, three-conductor cable. Buy TW No. 14 wire with red, black, and white insulation for wiring within the fixtures.

Wire one group of fixtures on the red wire and the other group on the black wire, and maintain the same coding throughout the room. The wiring within and between fixtures is the same as single-circuit wiring, except that you alternate fixtures on the black and red circuit. Otherwise, join fixtures and route cables as described on pages 76 and 77.

In a finished room you must add another two-conductor cable to a power source, install a switch, and route the cable through walls and ceilings, as described on pages 64 and 65. Use cable containing one red and one white conductor. Maintain red as the "hot" line throughout the second circuit. If red-and-white conductor cable is not readily available, color code the black wire in one circuit by coloring the insulation red at each connection.

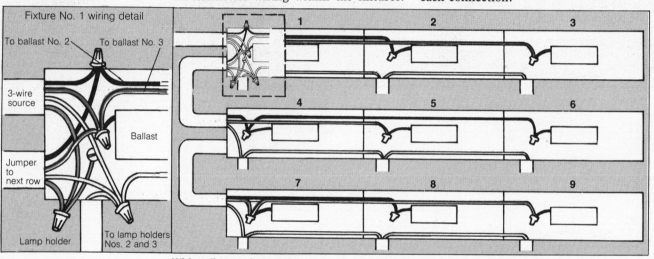

Wiring diagram for running three-wire power cable between rows of fluorescent fixtures so that half of them (Fixtures 2, 4, 6, and 8 on the black circuit) will be controlled by one wall switch and the other half (Fixtures 1, 3, 5, 7, and 9 on the red circuit) will be controlled by another wall switch. Insert shows detail of wiring for the first fixture.

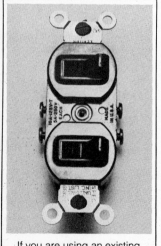

If you are using an existing switch box, use a double switch. Wiring is the same as for separate switches shown at right.

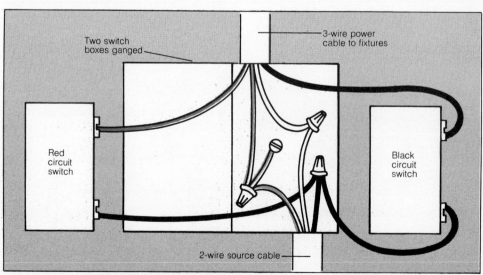

Wiring diagram for two-wire source cable and the three-wire power cable that leads to the fixtures, connected to two switches in two, gauged switch boxes. Connect the black wire from the source cable to two switches. Connect the black wire from the three-conductor cable to one switch, and connect the red wire in the three-conductor cable to the other switch. Wire the white wire and ground wire as shown.

Luminous ceilings

A luminous ceiling is the easiest type of ceiling to install in an unfinished area. The first step consists of mounting fluorescent fixtures, end-to-end, in evenly-spaced rows. Next, fasten hooks to the ceiling or to joists, and use wires to suspend metal runners from the hooks. Insert crosspieces between the runners to make equally square (2-by-2-foot) or rectangular (2-by-4-foot) openings. Finally, slip precut plastic panels—slightly larger than the square

Unfinished areas

When you want to maintain maximum overhead clearance, as in a finished basement for example, you can mount the fluorescent fixtures between floor joists and attach the plastic panels, runners, and crosspieces to the bottom edge of the joists. Forty-watt, single-lamp, rapid-start fixtures are best for this type of installation.

or rectangular openings—in place. They will be supported by the runners and crosspieces. These steps are shown in detail on the next two pages.

The various combinations of fluorescent-lamp types and plastic materials make it possible to achieve a wide variety of light levels and tints.

Pages 76 and 77 describe how to join fluorescent fixtures in rows and how to wire multiple rows.

The ballasts in fluorescent fixtures consume power and, therefore, give off heat. Take this into consideration when you plan how you will mount the fixtures. Make a mark on the outside of each fluorescent fixture to show the location of the ballast. Allow at least an inch or two of airspace above the ballast for heat dissipation.

What it takes

Approximate time: About two hours to install a luminous ceiling in a small area such as an entrance way or bathroom.

Tools and materials: You will need a pencil and ruler for marking old surfaces as you plan your ceiling layout. Hammer, bit and brace, and pliers will be needed for removing parts of old surfaces and fixtures. Materials you will be using are precut plastic panels, metal plastic-panel runners, 1-by-2-inch wood for side-support strips, round-head stove bolts (or wooden screws), and finishing nails.

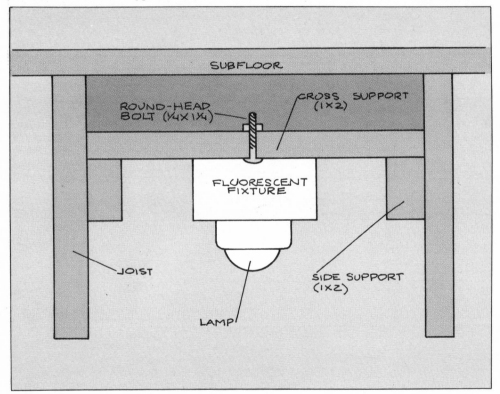

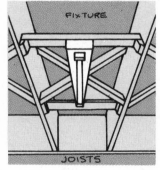

Measure the distance from the bottom edge of the floor joist to the subfloor above. A typical distance is 7¼ inches. Fluorescent lamps should be about 1½ inches above the plastic panels. The fixture (from the bottom edge of the lamp to the top surface of the fixture) measures about 3¾ inches. Mount 1-by-2-inch side-support strips along each joist between the bridging, as shown. (Do not remove the bridging.) Measure the exact distance between each of the joists (about 14½ inches) and cut the crosspieces to fit. Remove the lamps and cover from the fixtures. Then, use wood screws or round-head stove bolts to mount the crosspieces on the fluorescent channels. Hang the fixtures between the joists by tilting the fixtures and resting the crosspieces on the side-support strips. Do not fasten the crosspieces at this time. It will be easier to wire the fixtures if you can move them slightly. Wire the fixture for either one-circuit or two-circuit operation (pages 76 to 78). After the fixtures have been wired and tested, secure the crosspieces by driving finishing nails through the supports and into the joists at about a 45-degree angle. Once the fixtures are nailed in place, insert the lamps.

Installing ceiling support hardware and plastic panels

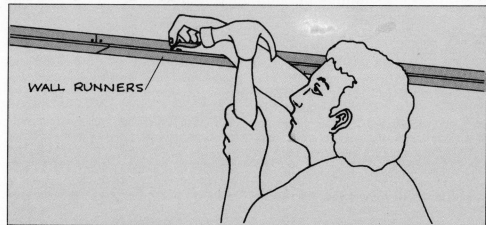

Wall runners are L-shaped pieces that are nailed directly to all four walls. Measure carefully, use a carpenter's level, and mark guidelines on the walls before you mount the runners.

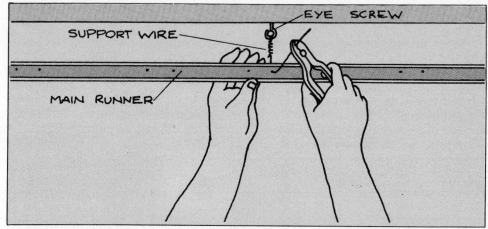

Install the main ceiling-panel support runners (T-shaped pieces) at right angles to the floor joists. Use support wires for this, as shown. Check the main runners with a carpenter's level and adjust the support wires as necessary. There may be a significant difference in the level of various floor joists.

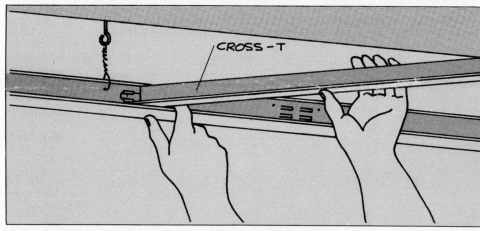

Install the cross-T pieces between the main runners and insert the panels. Open-mesh paneling or paneling with small perforations is best for between-joist installation. It allows for freer air flow. If you prefer solid paneling, install small sections of perforated or open-mesh paneling at each end of a joist run.

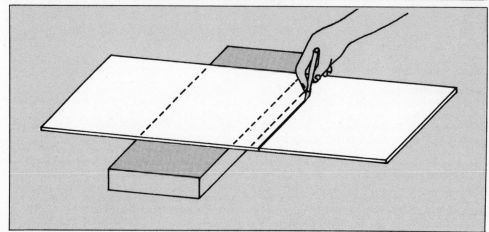

To finish areas that require panels of less than standard size, you will need an inexpensive plastic-cutting tool. Use this tool to score the smooth side of the plastic; then place a support under the panel along the scored line and break off the excess material.

Finished ceiling areas

When you install a luminous ceiling below a finished ceiling, determine the direction in which the ceiling joists run, and plan to mount the fixtures at right angles to them.

Locate each joist by drilling test holes or breaking away ceiling material (the existing ceiling will be concealed when you finish). Draw lines on the ceiling corresponding to the center of each joist. Take measurements and make a sketch of the ceiling, including the location of joists. If the space between the new, luminous ceiling and the original ceiling will be 1 foot or more, simply hang standard rapid-start fixtures on chains. These fixtures are available in 48-inch lengths with one or two lamps. The fixtures have built-in reflectors and come with hooks and chains for mounting. Use your ceiling sketch to plan locations, so you can screw the hooks into the joists.

Since standard center-to-center joist spacing is 16 inches, you will have at least three points to mount each 48-inch fixture through the ceiling to a stud.

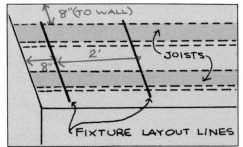

If there isn't enough room for hanging fixtures, use the following guidelines to plan your layout.
1. Unless an unusually high level of illumination is required, use 40-watt, rapid-start, single-lamp fixtures throughout.
2. Make a mark on the outside of each fixture to indicate the location of the ballast. Mount the fixtures so that there is at least 1 or 2 inches of airspace above the ballast.
3. Space the fixtures 8 to 12 inches from each wall.
4. Approximately 2 feet between rows of lights will provide good illumination.

Ruling the waves
You have to cut all those wires for suspending the main runners from the ceiling joists, right? Naturally, your suspended ceiling is lower by the length of the wires. Seems sensible, doesn't it, to cut all the wires the same length and attach them to the same spot on each joist? Don't you believe it. I get seasick every time I look up at my nice wavy ceiling. The darn floor joists were slightly different heights. So lay a level on the runners and adjust the wires *before* you attach the rest of the hardware and put in the panels.

Practical Pete

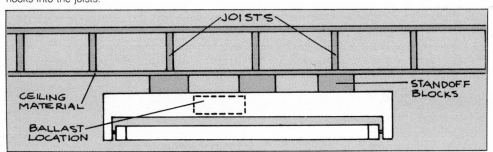

Cut pieces of 2-by-3-inch or 2-by-4-inch lumber, each about 4 inches long. Working on the 2-inch surface, drill a hole through the center of each piece. Use these blocks of wood as standoff mountings for the fixtures. Use 3-inch wood screws to attach the fixtures to the blocks. Insert a screw through a hole in the channel, then place a standoff block over the screw. Secure the fixture to the ceiling by putting the screw through the ceiling material and into a ceiling joist. Mount the L-shaped runners on the walls. Check these carefully with a carpenter's level. Secure the T-shaped main runners with hooks and wires. Check the runners with a carpenter's level at several points to ensure a level ceiling. You can easily make minor changes in runner height by adjusting the wire hanger. Snap crosspieces in place between runners, and insert plastic panels. Cut the panels as necessary.

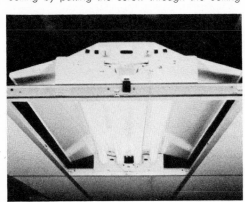

Complete two-lamp and four-lamp fixtures are available for installation in suspended ceilings. There are sizes for 2-by-2-foot and 2-by-4-foot panels. Make a ceiling layout and decide upon the fixture locations. Install cabling first. Route cabling to the location of each fixture. Secure cables to the ceiling with cable staples. Remove sheathing and

wire insulation at each location and let the cables hang. Install main runners and crosspieces as previously described. Mount and wire the fixtures in position, following the manufacturer's instructions. You can wire these fixtures for one-circuit or two-circuit control by following the wiring diagrams provided for ceiling-mounted fixtures.

8. TRACK LIGHTING

Track lighting is sophisticated and highly versatile. Originally, it was developed for use in museums and department stores. But in recent years, several manufacturers have adapted it for use in homes. Track lighting provides what decorators call accent light. It can be used to focus a narrow pinpoint of light on a single object or to flood large areas with light. In short, it can be used to create and to change the mood of a room.

Track lighting consists of an electrified track and matching fixtures—usually installed on the ceiling, 2 to 3 feet from the wall. Eight-foot sections can be mounted and joined at angles to follow the shape of a room. The track consists of continuous lengths of electrical conductors, mounted in plastic and enclosed in a metal channel. When specially designed fixtures are inserted in the track at any point, electrical connections are automatically made.

To make power connections, you can wire the track directly to a ceiling box, or attach a cord-and-plug adapter to the track and plug it into a wall outlet.

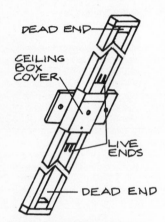

Center-feed components
These variations of live-end connectors accept track connections at both ends. You can use an existing ceiling box that happens to be in the middle of the run and install them like live-end connectors.

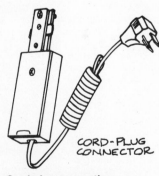

Cord-plug connection
Simply connect the end-piece adapter to the end of the track, plug the cord into a nearby outlet, and staple it to the ceiling and walls or conceal it behind drapes. The maximum electrical load on the track must not exceed 1000 watts.

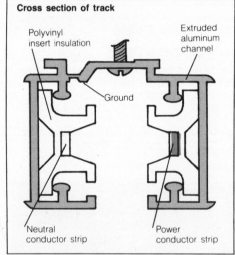

Cross section of track

Electrified track consists of an extruded aluminum channel that contains molded polyvinyl inserts. Three strips of metal are held in place by the polyvinyl. The strips of metal are connected to the power source. One strip (connected to the black power wire) is the "hot" line; another (connected to the white wire) is the neutral line; the third strip is connected to the ground wire. The polyvinyl insulates the aluminum channel from the conductors. When track sections are connected, the connectors make contact with the conductor strips in each section. In this way power is continued from section to section. The aluminum channel and the polyvinyl insert are designed to make accidental contact with the conductors almost impossible. Still, don't go poking around in there.

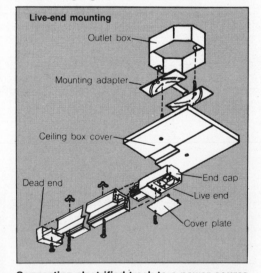

Live-end mounting

Connecting electrified track to a power source is easiest if you make the connection at a ceiling box. The fittings consist of a ceiling-box cover, an adapter that connects the cover to the ceiling box, and a "live" end piece that connects power to the track. To make the connection, first turn off source power to the ceiling box. Attach the cover to the box by means of the adapter. Feed the black, white, and ground power wires through the openings in the adapter and the cover. Then connect the power wires to coded, screw-type terminals in the live end piece. Attach the live end to its cover plate. When you connect the track to the live end, power will be applied to the metal track strips. Use end caps to close off the track. This will protect you from contact with the conductor strips at the end of the track.

Electrical rating

When connected directly to a ceiling or wall box, a track is rated at 20 amperes (2400 watts on a 120-volt line). The actual safe load, however, depends on the rating of the circuit to which the connection is made. If the circuit available at the ceiling or wall box is protected by a 15-ampere circuit breaker or fuse, then the track load on that circuit must not exceed 1800 watts.

If the circuit to which the track is connected also powers other lamps or appliances, the track load must be correspondingly reduced. For example, if the 15-ampere circuit mentioned above also supplied power to two wall outlets in which 150-watt lamps were plugged, the track load would have to be limited to 1500 watts. Different sections of the track can, of course, be connected to different circuits.

When track is connected with a cord-and-plug fitting, the maximum load is about 1000 watts. Make certain the circuit you plan to use can handle the additional load, however.

What it takes

Approximate time: Allow half an hour for each circuit wiring connection; 45 minutes for mounting and connecting each length of track; half an hour for positioning and mounting each lighting fixture.

Tools and materials: Track hardware, as described on pages 82 and 83, to fit your layout; solderless connectors, pliers, screwdriver, and voltage tester for circuit wiring; electric drill for mounting track. Most mounts include fasteners.

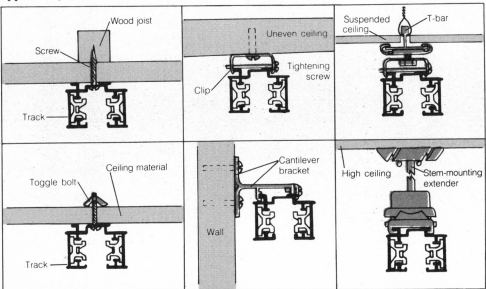

Track is made in 2-, 4-, and 8-foot lengths. Cut it to any length with a fine-toothed hacksaw. Remove burrs from the channel and conductors with a fine-toothed file; then brush any particles out of the track. Mount it directly onto joists with wood screws through holes or knockouts spaced along the track, or onto ceilings with toggle bolts.

There are many types of special mounts for use with track systems: clips for mounting track on uneven surfaces; extenders for lowering track; wall brackets; special fittings for attaching track to suspended ceilings.

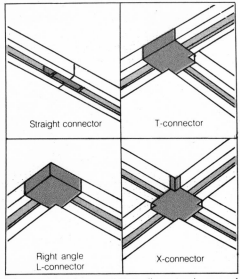

The variety of connectors allows track runs of almost any shape. In addition to straight connectors, right and left elbows provide right-angle, L-shaped track connections; T-connectors join three sections of track; X-connectors make four-way runs possible.

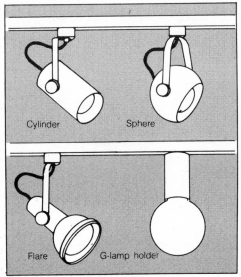

Typical track-lighting fixtures that swivel are cylindrical, spherical, and flare-end types. Lamps for the fixtures are available in 30- to 300-watt sizes. Standard general-service lamps, reflector lamps, and PAR (parabolic-aluminum-reflector) lamps can be used. G-lamp globes can be fitted to fixed holders that attach to track.

Planning hints:
While all track-lighting systems are essentially the same electrically, parts made by different manufactuers cannot be combined. Check the hardware available for track installation; the styles and types of fixtures available, and of course the cost.

9. SURFACE WIRING

What it takes

Approximate time: Surface wiring goes faster than adding outlets by fishing cable inside walls and installing new outlet boxes. Allow about two hours for running plastic strip wiring over a kitchen counter or workbench. Double the time for installing the same amount of baseboard surface wiring.

Tools and materials: There is little or no carpentry, plastering or painting required. To cut metal strip housing use a 40-tooth hacksaw blade. Plastic strip can be scored and broken with flat-end pliers (or a special trimming tool that cuts the housing and strips insulation off wires in one motion). Screwdriver, drill and bits, pliers, solderless connectors, the necessary lengths of track and associated hardware.

Surface wiring provides a safe, easy way to add outlets where you need them. As the name indicates, it is a way of adding outlets and fixtures to circuits without penetrating walls, ceilings, or floors. The wiring itself and the outlets and fixtures are all installed on wall or ceiling surfaces. There are two basic kinds of surface-wiring systems. One type consists of metal fittings and raceways (the channels that enclose the wiring); the other type is made of plastic. Metal raceways and fittings are grounded the same way armored cable is. Plastic raceways and fittings need a grounding conductor.

Surface wiring has some important limitations. It can be used only in protected, dry locations. Electrical codes in many areas limit surface wiring to one room; that is, a given surface circuit must begin and end in the same room. You can, of course, install another circuit in another room. The intent of this restriction is to keep surface-wiring raceways from penetrating walls or ceilings. Surface wiring is too conspicuous to be used in some areas. However, raceways and fittings are available in many finishes that blend well with surrounding surfaces. You can also help make surface wiring inconspicuous by keeping raceway runs low and horizontal. Vertical runs—unless concealed by drapes—are quite noticeable because the raceways cast shadows. Make vertical runs in or near corners or along door frames.

Metal and plastic raceway channels come in different sizes that hold various numbers of wires. The smallest size will easily accommodate three type-TW No. 14 or No. 12 wires. This should be large enough for most installations used for ordinary household appliances.

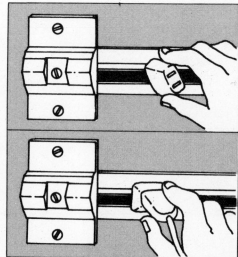

A plastic-surface wiring system has conductors embedded in a strip that is molded to accept specially designed outlets. The outlets are inserted in the strip at an angle and then twisted to a straight up-and-down position. This action makes an electrical connection between projections on the back of the outlet and the conductors in the strip.

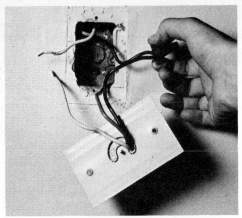

Power is connected to the plastic strips from an existing wall-outlet box. A special cover plate is provided to make electrical connections. The plate has two wires at the back and a fitting on the front. To install the plate, turn off power to the outlet. Remove the existing outlet faceplate, and disconnect and remove the receptacle.

Connect the wires on the cover plate to the power wires in the outlet box by means of solderless connectors. Then attach the cover plate to the box.

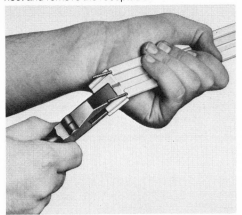

Trim the plastic strip at the ends to expose the conductors. To connect power to the strip, simply insert it in the fitting on the cover plate.

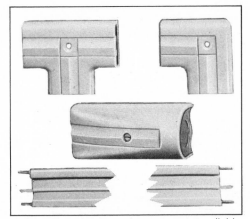

Elbows, Ts and right-angle corners are available, so you can route the strip wherever you want it. A special fitting is available to start the run from armored cable, if an outlet box cannot be used.

Metal raceways

Raceways are attached to walls in several ways. In one method, a two-piece channel is installed. The base piece is attached to the wall with nails or screws. Wires are routed along the base and then concealed by a cover that snaps into the base. In another, mounting clips are attached to the wall. Wires are fed through a rectangular tube held in place by the clips. An existing wall outlet usually provides power for the surface circuit. With power off, remove the faceplate and receptacle from the box. Then take the electrical connections from the receptacle. Mount an extension frame on the wall box. The extension frame has removable twist-out sections. Remove one and insert the raceway in the frame. Reattach the original electrical connections to the receptacle. Attach the wires for the surface circuit, routed through the channel, to the spare-receptacle terminals. Remember to connect black wires to brass-colored terminals and white wires to chrome terminals. Be sure to maintain the wire color coding throughout the surface circuit.

Mount the receptacle on the extension frame, then mount the faceplate.

Baseboard wiring

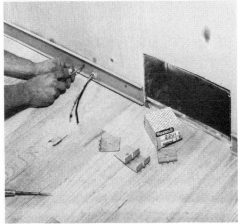

One type of metal surface is designed to simulate and replace baseboard. Outlets are installed directly in the raceway, eliminating the need for projecting boxes.

The metal channel has two sections. The rear section is attached to the wall in place of the original baseboard. Knockouts in the rear section can be removed, the power cable inserted, and secured to the raceway with a cable clamp. Use solderless

connectors to join the power wires to the raceway. The front section of the channel has openings for receptacles. The outlets are held in place by spring clips. Spacing of outlets can be varied to suit your needs. In some models an additional snap-on strip can be put on top of the power channel to conceal low-voltage wiring (hi-fi or intercom) or TV/FM antenna wires.

10. TROUBLESHOOTING APPLIANCES

What it takes

Approximate time: Varies with extent of disassembly. Anywhere from 15 minutes to an hour.

Tools and materials: A clean, well-lighted work area, set of Allen wrenches, set of Phillips-head and slotted screwdrivers, a container for small parts, a flashlight and magnifying glass.

Most of us have come to depend on appliances to do dozens of jobs around the house. And with proper use, they will perform for many years. The typical life span of a toaster, for example, is ten years. A refrigerator will last for 15 years; an electric range may last for 20 or more.

Appliances do break down, however. The next few pages will tell you how to prevent early breakdown and how to correct the majority of troubles that may arise in the life of an appliance.

When appliances suddenly stop

1. **Check the power.** Most kitchens have an appliance circuit with outlets near the work areas. When small appliances fail to work, check the circuit breaker or fuse for the appliance circuit. If several appliances—especially those that heat—are being used simultaneously and all stop working, a simple overload is probably the cause. A toaster and an electric frying pan on the same circuit can easily cause an overload, even on a 20-amp circuit.

When an appliance will not start, or if it stops working during operation, turn the switch off and unplug the cord as soon as possible. Plug in another appliance or use a voltage tester to check for power at the outlet. If power is OK, check the appliance cord and plug for damage.

For large appliances, plug in a work light or voltage tester to check for power at the outlet. If the appliance is wired directly to the service panel and the circuit breaker is not tripped or the fuse not blown, assume that power is reaching the appliance. If power circuits and cords are OK, go on to the next step.

Testing appliance cords

When you buy an appliance
- Large appliances often have instructions stapled to the inside of the carton. Check for them before the carton is discarded.
- Service contracts are often available from manufacturers and dealers. Ask for details.
- Keep an appliance file for all receipts, warranties, serial numbers, and operating manuals.
- Complete and return the purchase-information card.

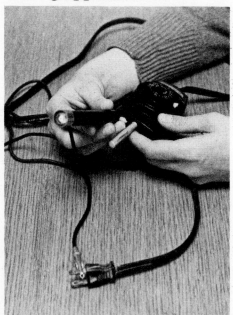

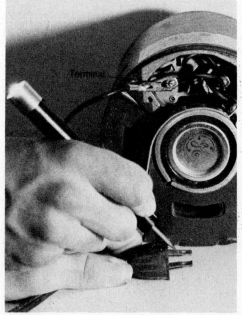

Small appliances. Remove the cord from the appliance. Using a continuity tester, connect the alligator clip to one prong of the cord plug. Insert the probe in each receptacle at the end of the cord that attaches to the appliance. The tester should light when the probe touches one—and only one—receptacle. Repeat the test with the alligator clip connected to the other plug prong. The continuity tester should light once on both tests. If it fails to do so on either test, there is a break in the cord. If, in either part of the test, the tester lights up when the probe is touched to *both* receptacles, the cord is shorted. In both cases, replace the cord.

Large appliances. Turn off power at the service panel. Unplug the appliance; then find and remove the access plate to the power source. (This will be located near the point where the power cord enters the appliance.) Behind the access plate, you will find two threaded terminals to which conductors in power cable are attached. Use your continuity tester to check from each plug prong to each terminal. The procedure and results should be the same as described for small-appliance cords.

2. Disassemble and inspect. The objective in fixing appliances is to inspect the workings with the least amount of disassembly. The procedure varies with different appliances, and may involve some detective work since main housing screws are often hidden. Typical methods of concealing screws are shown below.

The charts on page 90 will help you identify the type of motor generally used in the appliance you are troubleshooting. The motor characteristics column of the chart will help you locate trouble spots. **TIP:** For small appliances, first turn the appliance over and gently shake it. Loose parts will rattle or fall out.

Tips on locating screws

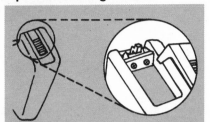

Snap-on parts. Check for removable plastic parts on the main housing. Mounting screws are often concealed underneath.

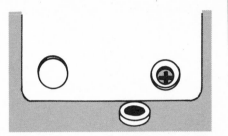

Felt, rubber, or plastic feet. Projections attached to the main housing to protect work surfaces may be glued over screws.

Deeply recessed screws. Don't pass up an opening in the housing because no screw is visible. Look in carefully with a flashlight. Screw heads are sometimes recessed an inch or more.

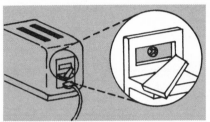

Metal inserts in plastic. The manufacturer's metal nameplate may actually conceal assembly screws. If you can, slip a knife blade between the edge of the metal insert and the plastic, and pry out the insert.

NOW you read the instructions She said something about the sewing machine not working right. So big-deal, fix-anything Pete does his macho bit and tears into the thing to find what's wrong. Kinda fun until—you guessed it—he finds he has a couple of strange-looking parts left over when he's got it back together. Big bill and no sympathy from the repairman. Turned out that the bobbin had not been inserted correctly and thread tension was turned a little high. Reading the manufacturer's instruction manual doesn't do much for the ego, but that's how repairmen make a lot of their money. *Practical Pete*

Disassembly techniques

Small motor-driven appliances. Housings usually support all internal parts—motor field, armature, bearings, gears, and so on. If you must take the unit completely apart, support the housing as you remove the main screws. Separate the housing parts slowly and carefully. This will reduce the chance of breakage and will also give you a chance to see how parts are positioned in the housing before they fall free.

Small heating appliances. These have no gears or bearings and generally have few small parts. Here, the trouble will most likely be due to a break in the heating circuit—the power cord, the heat control, or the heating element itself. You can locate the break by (1) disassembling the unit and testing the cord, as described on page 86; (2) checking the heat control (page 92); and (3) inspecting the heating element for breaks.

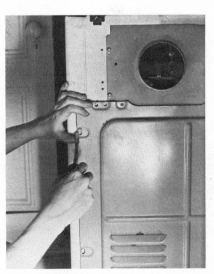

Large appliances. Remove front, side, or rear panels to expose the key parts. Rear panels are usually secured with accessible sheet-metal screws. For front and side panels, remove the screws or snap-out spring clips at the bottom edge. Bring the panel 2 or 3 inches from the appliance and push it up. This will free the panel at the top. Try removing the front panel first to expose the side-panel mounting screws.

Troubleshooting checklist

Many defects that appear electrical are not. You can save a lot of time and money by making sure you are following the operating instructions for the appliance. Manufacturers provide detailed instruction manuals for most appliances, especially those requiring delicate adjustments, like sewing machines. Since these vary from model to model, write to the manufacturer for the correct manual if you don't have one. Take the time to read the instructions for your model. It's tedious, but better than paying a repairman to "fix" your appliance by telling you how to operate it correctly. An iron that "spits" water, for example, is possibly doing so because of a faulty thermostat. More likely, though, the iron is at the wrong setting or the reservoir is too full. Before you decide to repair an item, make sure you're using it correctly.

Problem	Cause	Correction
Food Mixers and Blenders		
Will not run	Speed control dirty or defective	Clean or replace speed control.
Lacks power on all settings	Control switch defective	Replace switch.
Excessive noise and vibration	Brushes worn or chipped	Reshape or replace brushes (page 89).
Mixer heats up excessively	Gears misaligned or damaged	Realign or replace gears (page 89).
	Bearings worn or binding	Replace or lubricate bearings.
Can Opener		
Slow running	Cutting edge dull or chipped	Inspect. Sharpen or replace cutting edge.
Noisy	Gears defective	Disassemble and inspect gearbox. Lubricate if necessary. Replace broken gears.
Portable Fan		
Fan erratic or slow	Motor armature binding	Clean and lubricate motor bearings (page 91).
	If oscillating fan, oscillator gears broken or jammed	Check oscillator gearbox. Lubricate or replace broken gears.
Fan vibrates or "walks"	Blade unbalanced	Check blade for dirt accumulation. Clean as necessary. If blades are bent, realignment is difficult but worth a try.
Vacuum Cleaner		
Motor slow, little suction	Brushes worn or chipped	Reshape or replace brushes (page 89).
	Motor bearings worn or misaligned	Align, lubricate, or replace motor bearings.
	If cleaner has belt-driven brush, belt off drive pulley	Check belt. Reposition belt on drive pulley. If belt seems loose, it will jump out of pulley. Replace belt.
Motor sounds normal, poor suction	Vacuum leak	Check hose for clogging or breaks. Check attachments for tight fit at joints. Hose may be temporarily patched with plastic tape. Worn attachments should be replaced.
	Exhaust outlet blocked	Check exhaust filter. Clean or replace.
Sewing Machine		
Machine slow and noisy	Lubrication required	Check instructions and lubricate as indicated.
Waffle Iron		
Too little or too much heat	Defective thermostat	Check thermostat. Clean if necessary. If temperature is still off, replace thermostat.
Waffles stick to grill	Grill not seasoned	Heat for 30 minutes after brushing grids with cooking oil.

Problem	Cause	Correction
Toaster		
Toasts one side only	One heating element open	Repair or replace broken element.
Toast does not pop up	Bread caught in carriage wires	Unplug toaster. Remove bread. Shake out crumbs.
	Pop-up spring broken	Check spring. Replace if broken.
	Hold-down latch caught or binding	Check latch mechanism. Clean, straighten, or replace.
Toast too light or too dark	Linkage from color control to release mechanism broken or loose	Check linkage. Make sure sliding parts are properly engaged. Replace if broken.
Coffee Maker		
Water gets warm, but doesn't perk	Defective thermostat	Replace thermostat—the circular unit located in base.
Steam Iron		
No steam	Steam ports clogged by mineral deposits	Clean with small brush dipped in vinegar.
Iron too hot or not hot enough	Thermostat incorrectly set or defective	Check and correct thermostat setting. If temperature is still wrong, disassemble and check thermostat. Have repair shop replace if defective.
Automatic Washer		
Tub does not fill	Water hoses disconnected or blocked	Connect hoses. Check for kinks or pinching.
	Shutoff valve closed somewhere in water-supply line	Disconnect hose at supply. Check water flow. Open valves as necessary.
Tub does not drain	Drain hose blocked	Check drain hose for kinks or pinching. Remove kinks and reroute hose as needed.
No spin cycle	Drive belt loose or broken	If belt is loose, tighten; if stretched or broken, replace.
Washer vibrates and "walks"	Small or uneven load distribution	Load must be large enough for even distribution.
Electric Clothes Dryer		
Does not start	Defective door interlock switch	Replace door interlock switch.
Does not heat	Defective heating element	Disassemble dryer and check heating element. Repair or replace as indicated.
Drum does not rotate	Drum binding or drive belt broken	Check drum for free movement. Small items of clothing can work between drum and housing. Disassemble and clear. Replace belt if broken.
Automatic Dishwasher		
Does not fill	Supply line turned off or blocked	Check supply line.
Dishes not clean	Water not hot enough	Check domestic water temperature. If heating system is *instantaneous*, be sure no other hot water is used during fill and rinse cycles.
Does not drain	Drain hose blocked	Clear or unkink as needed.
Refrigerator		
Interior temperature not cold enough	Control setting too low or control defective	Adjust control. If still not cold enough, check other items below. Have control replaced if defective.
	Door seal worn out	Replace door gasket if brittle, split, or worn.
	Inadequate ventilation around vents	Check for and remove obstructions.
	Defrost cycle on continuously	Timer defective. Have timer replaced.

Small motors

Appliance	Motor type	Characteristics
Food mixers Blenders Vacuum cleaners Sewing machines	Universal	High power. AC or DC. Wide range of speeds. Has commutator and brushes. Current flow is from one power line through one of the flat-wound, stationary field coils to one brush, through rotating armature, through other brush to second flat field coil to other side of power line.
Fans Can openers Hair dryers	Shaded pole	Light-duty motor. AC only. No commutator or brushes. Current is supplied to stationary field coil only. Small and compact.
Rotating cordless devices	Permanent magnet	Similar to universal motor except field coil is permanent-magnet type and requires no power. Brushes and commutator apply power to rotating armature. Low power output and consumption. Usually designed to run from rechargeable batteries.
Clocks Turntables Tape decks	Synchronous	Constant speed linked to 60 Hertz AC. Power is applied to stationary field. No brushes. Little maintenance required. Suited to any light-load application requiring constant speed.

Large motors

Appliance	Motor type	Characteristics
Washing machines Grinders Saws Lathes	Split phase (1/3 hp or less)	Operates on 60 Hertz AC. No commutator or brushes. Two sets of windings—one for starting; one for running. Centrifugal switch cuts out power to the start-winding when motor reaches running speed. Motor enclosed in own housing.
Large fans Water pumps Heavy shop tools	Capacitor start (up to 10 hp)	Operates on 60 Hertz AC. Capacitor provides power "kick" to start heavy loads. Centrifugal switch cuts out capacitor circuit when motor reaches running speed. Motor enclosed in own housing; capacitor mounted on top.

Universal motor

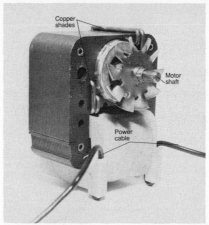

Shaded-pole motor

Permanent-magnet motor

Synchronous motor

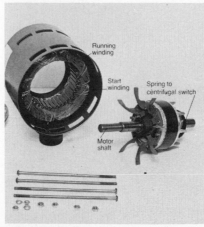

Split-phase motor

Capacitor-start motor

Motor parts: function and malfunction

Commutator and brushes supply power to the rotating part (armature) of a motor. The commutator is a segmented metal cylinder separated by insulating material. Brushes are small blocks of carbon that remain stationary and press against the commutator as it rotates. This causes current to flow between them.

If an appliance is sluggish, check brush contact by running the appliance in a dimly lighted area. Sparking will be visible through the ventilation openings in the appliance.

Small bluish sparks where commutator and brush meet are normal. Light bright sparks or bright and dim loss of power. You can shape new or chipped brushes to make good contact in two easy steps. First, wrap a piece of fine-grit sandpaper (rough side out) around the commutator. Then insert the brushes in the holder so they press against the sandpaper. Rotate the commutator back and forth by hand until the brush ends are shaped to fit.

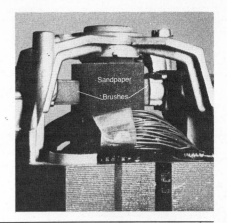

Centrifugal switches are used on split-phase and capacitor-start motors that have one circuit to start the motor and another to keep it running. The part that switches it from one circuit to the other is a spring-loaded centrifugal switch.

Centrifugal-switch contacts are normally closed when the motor is at rest. As the motor starts to turn and build up speed, centrifugal force moves the flyweights outward. This causes the arm to turn on the pivot, removing pressure from the switch and allowing the contacts to open. **Tip:** If switch contacts become clogged or dirty, the switch mechanism will stick and the *motor will not start.* To clean, remove the motor end bell and run a fine-toothed file between switch contacts to clean them. Clean mechanical parts of switch with small brush dipped in any household solvent. Dab light oil on sliding surfaces and pivot points. Hand operate parts to ensure free movement. Badly bent or broken switch parts will have to be replaced.

Gears convert motion from the motor to the part which is being driven and may be made of either metal or plastic. Metal gears are usually heavily lubricated; plastic gears are not. If an appliance motor hums but the appliance runs slowly, noisily, or doesn't run at all, the gears may be faulty. Gears can jam if they suffer mechanical shock (as by dropping or by severe overloading). Jammed gears can be repaired; broken or worn gears cannot. To repair faulty gears, separate them with a wooden or plastic stick. Rotate gears by hand to check for free movement. If they don't move freely, the gearing or the appliance must be replaced.

Motor bearings reduce friction between a fixed and moving part. If worn, dry, or misaligned they can cause poor performance and motor burnout. If the motor is permanently lubricated, make no attempt to add more. If oil is required, check for clogged wicks at lubrication points. Pull wick out of holder with tweezers and clean it in dry-cleaning solvent. Replace it when dry and add specified amount and type of lubricant.

When motors overheat

Heat can be destructive to any motor. Small appliances often have fan blades attached to one end of the motor armature to force air through the housing. Make sure ventilation openings are not blocked.

Most appliances are not intended for continuous use. Motor bearings will overheat and may bind or sieze, causing severe motor damage. Units should be turned off and allowed to cool any time the housing becomes too hot.

Heating elements

Heating coil

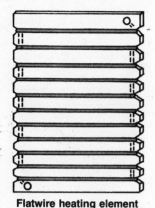

Sheathed heating coil

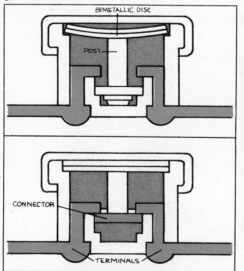

Element	Appliance	Characteristics
Heating coil	Waffle iron	Spiral coil of heat-producing wire mounted on ceramic standoffs. Used in appliances where coil heats a surface that in turn cooks food. Breaks in wire are easy to spot by inspection.
Sheathed heating coil	Broiler	Heating coil enclosed in a ceramic-lined, steel tube that protects coil from grease. Heats food directly. Breaks not visible. Continuity check must be made.
Flatwire heating element	Toaster	Resistance wire wound on flat mica insulator. Spacing of windings can be varied to produce even heat. Breaks can be found by careful visual inspection.

Flatwire heating element

When buying appliance parts

Whether you buy parts from a local dealer or shop by mail order, it is essential to provide full identification of the part you need.

Manufacturers and dealers stock parts by part number. Inspect the defective part carefully to find the number marked on it.

If none can be found, chances are the part is not available separately and you will have to purchase the smallest assembly in which the part is located.

Find the part number on the larger assembly. Record all numbers marked on this part, to be sure to have the right one. Include the appliance model number as it appears on the nameplate. Include dash numbers or letters added to the model number, as this may represent part changes.

Be wary of parts described as "just as good" or "better" than the manufacturer's original part. These parts may in fact be better made, but using them may cause you to replace other parts or modify the appliance. Accept these parts only if nothing else is available, and you are assured of a full refund if they can't be used.

Temperature control in appliances that heat

The basic element in controlling heat is a bimetallic unit that consists of two different metals fused together. Both expand when heated. But one expands faster than the other, causing the bimetallic unit to warp or bend. This, in turn, opens and closes electrical contacts, turning current on and off.

Electrical heating appliances are always either on or off. The heat level is controlled by varying the time between the on and off stages. If the current is on more than off, a relatively high temperature is maintained. If it is off more than on, a lower average temperature results.

There are two types of heat controls in general use:

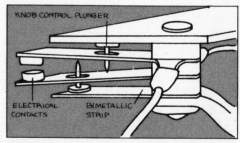

Fixed heat controls, like those on coffee makers, have a factory-set temperature. They are often operated by a circular bimetallic unit. When the desired temperature is reached, the bimetallic disc bends, pushing a plunger that opens the switch contacts. As temperature drops, the disc straightens, closing the contacts. Thus, an average temperature is maintained.

Repair work on this type control is limited. If the unit is faulty, all you can do is clean the switch contacts. To do this, press contacts together with a piece of soft cloth or a dollar bill between them gently sliding cloth back and forth.

(Note: On many fixed-heat appliances, the heat-control housing is sealed. In this case, when something goes wrong, the complete heat-control unit must be replaced.)

Adjustable heat controls are found on units like toasters and electric frying pans. On these units, a temperature-control knob activates a plunger that applies pressure to one of the electrical contact arms. When heated, the bimetallic unit—usually in strip form—applies pressure to the other arm. The less pressure applied by the control knob, the more pressure required by the bimetallic strip to separate the contact arms and switch off current.

Adjustable temperature appliances are fairly easy to repair, as the control-knob unit is accessible with some disassembly. The most common problem is uneven operation, caused by bits of food lodged between the contact arms and the knob or plunger. To fix, brush or scrape out food particles, and lightly rub the contact points with cloth or paper.

Room air conditioners

Wiring

Before buying a unit, know the amperage of the outlet it will be plugged into. Be sure installation will not overload existing circuits. Most home and apartment circuits are wired for a minimum of 115 volts/15 amps, although utility rooms (laundries and kitchens) may be wired for 20 amps to handle the load of large appliances.

Units rated at 12,000 BTUs or more are designed to operate on 208 or 230 volts and, like electric dryers, require special circuits. However, if circuit capacity is not over-taxed, you may be able to accommodate a unit of a maximum of 14,000 BTUs on a 120-volt circuit. Check with your local utility company for precise information on adequate circuitry.

Never remove the third prong from your unit's plug. Two-hole adapters known as "cheaters" are not recommended because they may not provide adequate grounding for the electrical charge. If your home has only two-prong outlets, install one that accepts a three-prong air-conditioner plug.

15 Amps 125 Volts

20 Amps 125 Volts

15 Amps 250 Volts

20 Amps 250 Volts

30 Amps 250 Volts

Seasonal maintenance

Wash air filters—located behind removable front panel—in warm, sudsy water, dry thoroughly, and replace. Some units require complete filter replacement when dirty; others are made of a metal mesh which can be cleaned and coated with a special dirt-catching aerosol. Check owner's manual on filter care. Wash every two to four weeks, depending on air quality of your area.

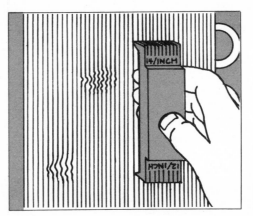

Vacuum coil fins with radiator attachment to clear away dirt, or clean with a soft brush. Straighten bent coil fins with fin comb available at refrigerator-parts stores. Spacing of teeth should match coils on unit. Coil fins are aluminum and bend out of shape from vibration and severe temperature changes due to ice forming on coils.

Tip: Exterior metal housing, though usually treated with rust- and corrosion-resistant materials, can become unsightly from severe exposure to the elements. You can touch up and further protect by wire-brushing rusted areas down to the metal finish and smoothing the surface with medium-grade sandpaper, feathering the edges as you go. A coat of metal primer and a finish coat of metal paint will protect it for years to come.

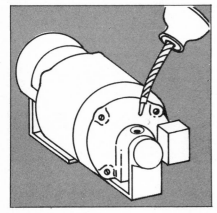

Oil fan bearings and condenser motor with SAE #10 or #20 nondetergent motor oil or type specified by manufacturer. Three drops on each spot will do it. Straighten bent fan blades. Check for refrigerant leaks—oily looking deposits around copper tubing. Check evaporator and condenser fan and motor belts for slippage. Replace or adjust by tightening.

Check mounting screws, nuts, and bolts which may have worked loose from vibration. Periodically tighten those on mounting frame. Also check screw mountings on mechanical moving parts inside unit. Some will need slight adjustments, others, like the compressor, should float in their mountings, and overtightening hold-down bolts causes noisy operation. Check owner's manual.

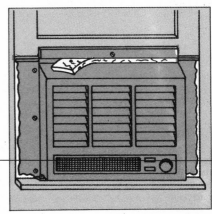

Replace loose or brittle window seals (weather stripping). Check for defective caulking on through-wall units. Air leaks reduce efficiency. Trim outside shrubbery to allow sufficient air flow.

Index

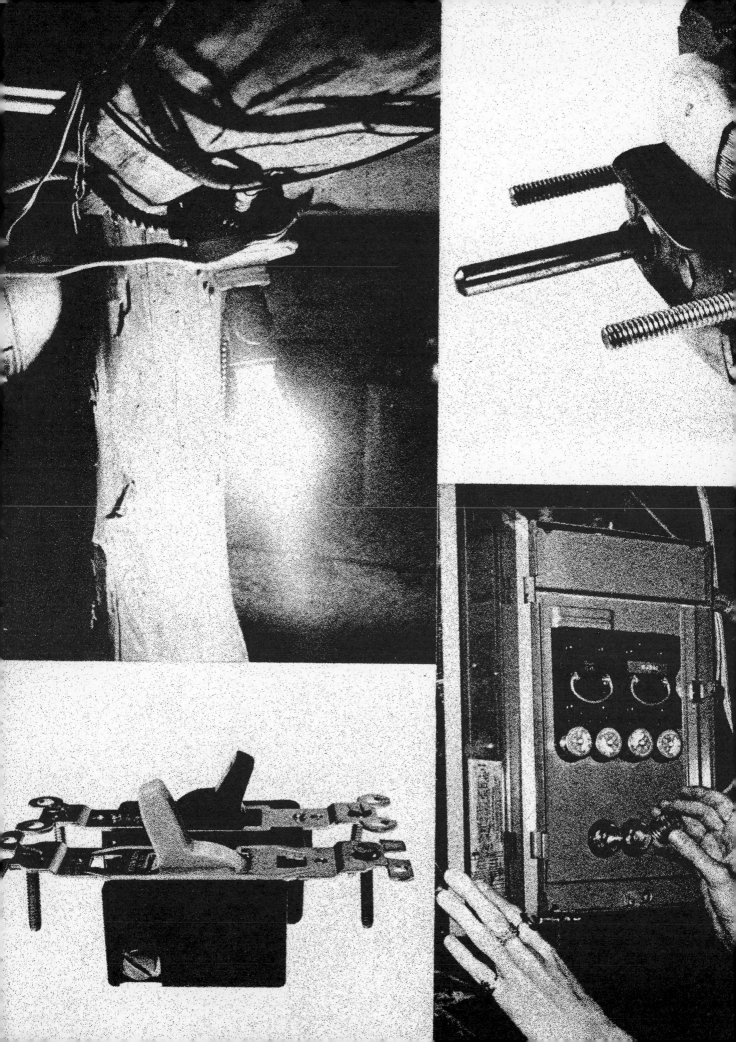